DIGITAL
VIDEO BASICS

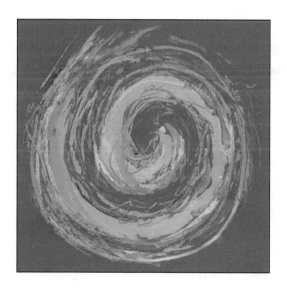

by Scott Schaefermeyer

THOMSON

COURSE TECHNOLOGY ™

Australia • Canada • Mexico • Singapore • Spain • United Kingdom • United States

Digital Video Basics
is published by Course Technology.

Author
Scott Schaefermeyer

Managing Editor
Donna Gridley

Product Manager
Allison O'Meara

Editorial Assistant
Amanda Lyons

Content Project Manager
Matt Hutchinson

Marketing Manager
Tiffany Hodes

Quality Assurance Testers
Susan Whalen, Serge Palladino,
John Freitas

**Manuscript Quality Assurance
Lead**
Jeff Schwartz

Developmental Editor
Cheryl Morse

Composition
GEX Publishing Services

Art Director
Kun-Tee Chang

How to Use This Book

What makes a good computer instructional text? Sound pedagogy and the most current, complete materials. Not only will you find an inviting layout, but also many features to enhance learning.

Objectives—
Objectives are listed at the beginning of each lesson, along with a suggested amount of time to complete the lesson. This allows you to look ahead to what you will be learning and to pace your work.

Step-by-Step Exercises—Preceded by a short topic discussion, these exercises are the "hands-on practice: part of the lesson. Simply follow the steps, either using a data file or creating a file from scratch. Each lesson is a series of step-by-step exercises.

Marginal Boxes—
These boxes provide additional information about the topic of the lesson.

Vocabulary—
Important terms are identified in boldface throughout the lesson and summarized at the end.

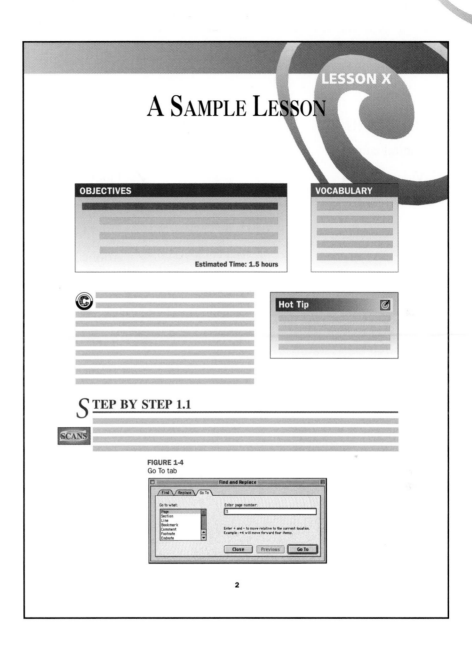

Enhanced Screen Shots—Screen shots now come to life on each page with color and depth.

Summary—At the end of each lesson is a summary that prepares you for the end-of-lesson activities.

Vocabulary/Review Questions—Review material at the end of each lesson and unit prepares you to assess the content presented.

Lesson Projects—
End-of-lesson hands-on application of what has been learned

in the lesson allows you to apply the techniques covered.

Critical Thinking Activities—Each lesson gives you an opportunity to apply creative analysis and use various resources to solve problems.

START-UP CHECKLIST

HARDWARE

✓ Digital camera

✓ Video camera

SOFTWARE

✓ Microsoft Office: Word, Excel

✓ Final Draft

✓ Final Draft AV

✓ Final Cut Express HD

✓ Adobe Premiere Elements 3.5

TABLE OF CONTENTS

UNIT 1 PREPRODUCTION

UNIT 2 PRODUCTION

UNIT 3 POSTPRODUCTION

TEACHING AND LEARNING RESOURCES FOR THIS BOOK

Instructor Resources CD

The *Instructor Resources CD* contains the following teaching resources:

- The Data and Solution files for this course.

- ExamView® tests for each lesson. ExamView is a powerful testing software package that allows instructors to create and administer printed, computer (LAN-based), and Internet exams.

- Instructor's Manual that includes lecture notes for each lesson, answers to the lesson and unit review questions, references to the solutions for Step-by-Step exercises, end-of-lesson activities, and Unit Review projects.

- Copies of the figures that appear in the student text.

- Suggested Syllabus with block, two quarter, and 18-week schedule

- PowerPoint presentations for each lesson.

ExamView®

ExamView is a powerful objective-based test generator that enables you to create paper, LAN, or Web-based tests from test banks designed specifically for your Course Technology text. Utilize the ultra-efficient QuickTest Wizard to create tests in less than five minutes by taking advantage of Course Technology's question banks, or customize your own exams from scratch.

PREPRODUCTION

Unit 1

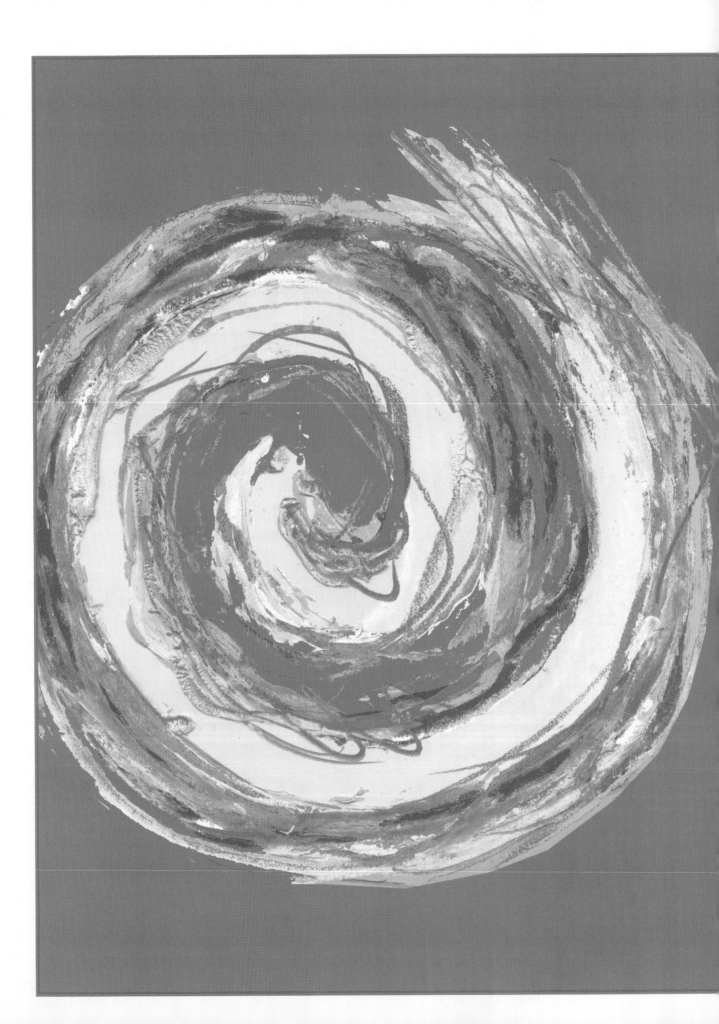

VIDEO AND DIGITAL VIDEO BASICS

OBJECTIVES

Upon completion of this lesson, you should be able to:

■ Explain persistence of vision

■ Explain scanning

■ Describe a cathode ray tube

■ Describe the difference between interlaced and progressive scanning

■ Explain the use of fields and frames

■ Explain time code and how it is used

■ Explain the analog video signal

■ Describe digital video sampling

■ Understand the concept of video formats

■ Explain the differences between DV and other digital formats

■ Understand the concepts of compression and color sampling

Estimated Time: 1.5 hours

VOCABULARY

Analog

Bandwidth

Capture card

Cathode ray tube (CRT)

Chrominance

Codec

Color sampling

Compressed

Field

FireWire

Frame

Frame accurate editing

Frequency

High definition (HD)

Hue

Interlaced video

Linear

Luminance

National Television System Committee (NTSC)

Noise

Nonlinear

Persistence of vision

Pixels

Progressive scanning

Red, green, blue (RGB)

Resolution

Sampling

Introduction

The first "movie" I ever made was with a borrowed 8 mm camera. I did not have any idea what I was doing, but when the film came back from processing, I sat down at the kitchen table with a razor blade. I held the roll of film up to the light, looked at each frame, and decided which frames to keep and which ones to cut out. I finished by taping the final edit together with a big roll of tape. It took forever, but when I was finished, I was hooked.

I was also in trouble. I cut the kitchen table to pieces and jammed my dad's projector with tape.

A couple of years later, my dad let me edit a high school project on a video-editing station at his office. It was a VHS cuts-only system (meaning the only transitions were straight cuts), but it was a lot faster than a razor blade and tape. My teacher and class were amazed at how professional my project looked, but by today's standards, it was awful.

When I first started working in professional video production, the editing equipment alone filled an entire room and cost hundreds of thousands of dollars. Today, anybody can buy a decent camera and editing system for less than $2000 and could probably find a workable system for less than $1500.

What has caused the cost of video production equipment to drop so dramatically? A lot of reasons have contributed to this decrease, but one of the most important is digital video.

Introduction to Video

Before you can work with video, you need to understand its basic principles, explained in the following sections.

Persistence of Vision

Start off by grabbing 16 pieces of scratch paper and a pencil. Number the pages from 1 to 16 in the upper-left corner. Now draw a small circle in the bottom-right corner of page 1 so that the bottom of the circle is at the paper's bottom edge, and the right side is touching the paper's right edge. It should look similar to Figure 1-1.

FIGURE 1-1
Your page 1 drawing

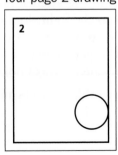

On page 2, draw the same size circle along the right edge but a little way up from the bottom of the paper. Try to make your drawing look like Figure 1-2.

FIGURE 1-2
Your page 2 drawing

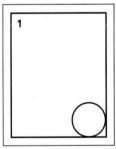

Draw or trace that same circle at the right edge of the remaining pages, but draw the circle a little farther from the bottom on pages one through eight. For example, the circle on page 3 should be a little higher up than on page 2, the circle on page 4 should be a little higher up than on page 3, and so on. Then, on page 9, start from the bottom-right again and repeat the process. When you are finished, put the papers side by side so that they look like Figure 1-3.

FIGURE 1-3
Your 16 pages

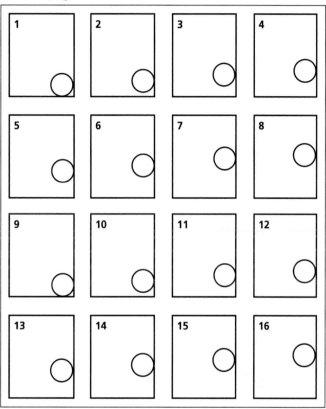

Now stack the pieces of paper in order so that page 1 is on top and page 16 is on the bottom. Flip through the pages with your thumb. If you did it right, the ball appears to move up and down on the page as you flip the pieces of paper.

Is the circle moving? No, but it appears to. Most film and video teachers teach that the illusion of movement is caused by **persistence of vision**; this theory states that the human eye holds each still image for a fraction of a second, with the image remaining on the retina long enough to blend the image with the next one. The result is that the combined images create the illusion of smooth motion. Any film student learns this theory, but some scientists reject the idea that it has anything to do with watching a motion picture.

Whether it is because of persistence of vision or another reason, however, the circle appears to be moving. Film displays 24 still images each second (also called frames per second, or fps) with the subject in a slightly different position in each image. NTSC video displays 30 frames each second (as explained in the next section); other video standards use other frame rates. So why is 24 fps used for film? The human eye can perceive motion at lower frame rates, but in the theater, images flicker at lower rates. The 24 fps rate creates the illusion of motion and removes the flicker.

Scanning

To project an image onto the screen, a film projector uses a powerful bulb to shine light through the film. The screen and film are in the same location. Television images, however, are sent all over the world and cannot rely on a powerful light bulb. Television technology has been based on the cathode ray tube (CRT) for almost 100 years. Newer technologies exist, but they are based on the same idea. You can find a simple CRT in a black-and-white TV, so I have used that as an example. A CRT has two basic parts, shown in Figure 1-4: an electron gun in the back and a screen in the front.

FIGURE 1-4
A basic CRT

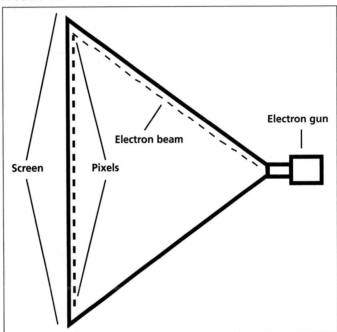

The inside of the screen is covered with thousands of tiny phosphorus dots called pixels. The electron gun receives the input signal and shoots an electron beam at the screen, lighting up the pixels. The stronger the beam, the more brightly the dots light up. If the beam is strong, the pixel is white. If the beam is not on at all, the pixel is black. If the beam is somewhere in between, the pixel is gray.

The back of a color CRT holds three electron guns: red, green, blue (RGB). The pixels on the screen are also RGB. The red beam hits the red pixels, the green beam hits the green pixels, and the blue beam hits the blue pixels. When the TV needs a white spot on the screen, all three electron guns fire at the same time. When the picture requires a black dot, none of them fire. All the other colors on the screen use varying combinations of red, green, and blue at different intensities of each color.

A single pixel, however, doesn't make an image. The pixels are placed in lines that extend across the screen. National Television System Committee (NTSC) video, the standard for American video, is made up of 525 of these lines stacked top to bottom, which are called scan lines. The number of scan lines determines the image's resolution or how much detail is in the image. More scan lines means more pixels, and more pixels means higher resolution and more detail. The electron beam starts in the top-left corner of the screen (from the viewer's point of view) and moves horizontally to the right, lighting up each pixel as it moves. When the beam reaches the right edge of the screen, it jumps back to the left to scan another line. Instead of scanning the second line, however, it scans horizontally across the third line. When the beam finishes the third line, it moves back to the left again, scans horizontally across the fifth line, and continues down the screen scanning the odd scan lines (see Figure 1-5).

FIGURE 1-5
Pattern of odd scan lines on a TV screen from the first field

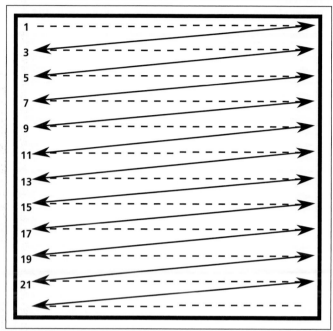

After the odd lines have been scanned, the television has displayed half a frame, which is called a **field**. The electron gun then moves back to the top left of the screen and scans the even lines, as shown in Figure 1-6.

FIGURE 1-6
Pattern of even scan lines on a TV screen from the second field

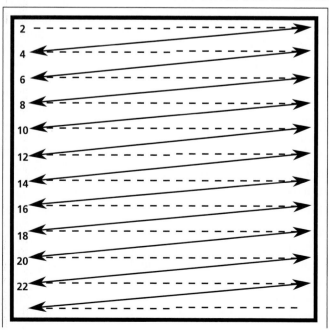

It takes 1/60 of a second for the television to scan the first field and another 1/60 of a second to scan the second field. These two fields fit together to display a complete **frame** (see Figure 1-7). Using two fields to create a complete image is called **interlaced video**. NTSC video runs at 30 frames per second (30 fps) or 60 fields per second.

FIGURE 1-7
The two fields interlace to produce a complete frame

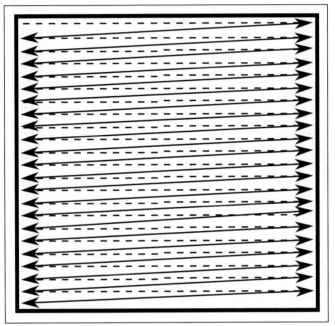

So why does video need interlacing fields when film does not? When you are watching a film in a theater, the actual images are right there, close to you—at least they are in the same building. Video, however, is a signal, not a physical image you can hold up to the light and see. The signal is transmitted over long distances through the air or across wires. These delivery methods have a limited **bandwidth**, which you can think of as the size of a pipe that information travels through. You do not need a lot of bandwidth (or a huge information pipe) if you split the information into pieces and send the pieces one after another.

An advantage film has over video is that a film frame is a complete image, compared to the two half-images that make up a video frame. The pixels used to create the first field of video start to fade by the time the second field is scanned, so a television image is not as crisp or detailed as a film image. A film image is crisper, with more detail and more vivid colors.

Progressive scanning, one of the advances in television technology, works much the same way as interlaced scanning, but instead of displaying a frame in two fields, it creates each frame by scanning the scan lines in order: 1, 2, 3, 4, 5, and so on (see Figure 1-8). This type of scanning creates a cleaner, crisper image. ATSC video can be

Note

NTSC created the standards used in American video in the 1940s and 1950s. In the early days of television, different manufacturers wanted their systems to be used, so NTSC was formed to examine the options and create a standard. NTSC is based on compromises, which, like all compromises, do not address every need but meet the majority. NTSC standards include the use of 525 scan lines and the interlaced 29.97 frame rate. (You might be wondering "Why 29.97?" You learn why in the "Time Codes" section.) The Advanced Television Systems Committee (ATSC) created higher-quality television standards, but they have not been implemented completely yet.

displayed at 30 fps but also allows different frame rates for progressive scan video, such as 24 fps and 60 fps. **High definition (HD)** is another product of ATSC. True HD uses 720 progressively scanned lines (known as 720p) or 1080 interlaced lines (1080i). The 720 lines add more detail in action scenes and 1080i produces more detail overall but is not as clean on fast action scenes. ATSC video is being phased in, but the NTSC standard is still in use and is the most common standard.

FIGURE 1-8
Progressive scanning scans every line in order

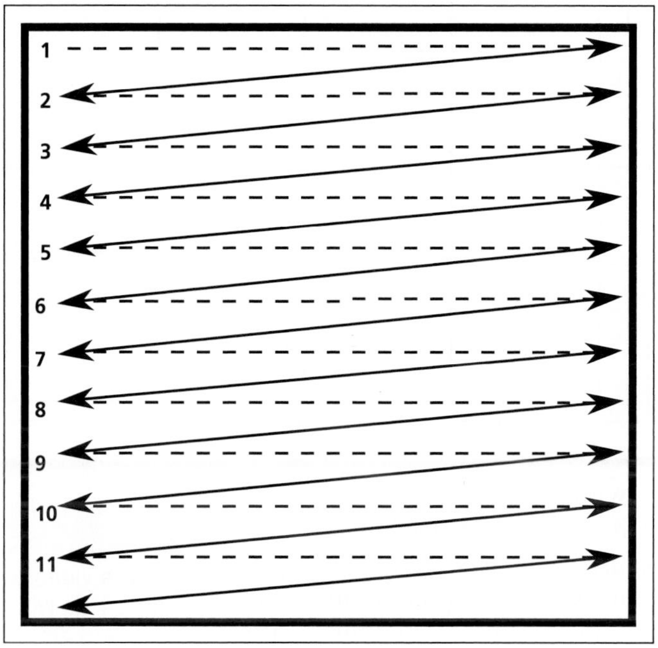

Time Code

To make working with video efficient and effective, each frame has its own unique identity. Instead of a name, however, each frame has a number. This identifying number is called a **time code** and looks something like this: 14:54:32:12. In this example, 14 represents hours, 54 means minutes, 32 means seconds, and 12 is the frame number. Because the first three numbers are based on time of day, you never see a number higher than 23 for the first number because there are only 24 hours in a day. Similarly, with 60 minutes in an hour and 60 seconds in a minute, you never see a number higher than 59 for the second or third number. Finally, you never see a number higher than 29 for the fourth number because there are only 30 frames of video per second.

With film editing, you can physically handle each frame and cut at the frame you want. Video, however, is an electronic signal. You can see the frame, but you cannot physically touch it. With time code, however, editing systems can identify every frame, and video editors can specify exactly which frames to include and which frames to cut. This type of editing made possible by time code is called **frame accurate editing**; without it, a clean, professional edit is nearly impossible.

Aspect Ratio and Resolution

Standard definition NTSC video is made up of 525 scan lines, but only about 480 lines make up the visible picture. The remaining 45 lines or so are used to carry information other than image information, such as time code and closed captioning. If the visible picture is 480 pixels tall, how wide is it? The width is determined by the **aspect ratio**. The aspect ratio for standard definition television is 4:3, which means that for every four pixels across, you have three pixels up, as shown in Figure 1-9.

FIGURE 1-9
Standard definition TV uses a 4:3 aspect ratio

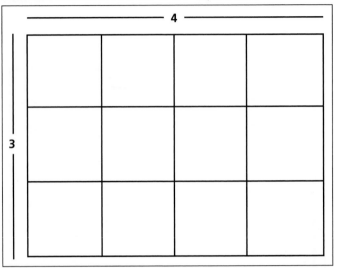

Using this 4:3 ratio, with eight pixels across, you have six pixels up; similarly, if you have 24 pixels across, you have 18 pixels up. Because computers are used so much in video, thinking of frame size in computer terms is easier. The aspect ratio for most computer monitors is also 4:3. If you have 480 visible pixels up on a computer monitor, you have 640 pixels across. (To do the math yourself, divide the number of visible pixels up—480—by the second number in the aspect ratio, which is the 3 in 4:3. You come up with 160. Multiply 160 by the first number in the aspect ratio, which is the 4 in 4:3, and you get 640.)

This explanation of aspect ratios is not entirely accurate in terms of a television image, but for the purposes of this discussion, it is close enough. The TV actually determines the number of pixels across. Some TVs have fewer than 640 pixels across, but the pixels are spaced farther apart to make it seem as though they have 640 pixels. That is why some TVs have a clear image, and others, with fewer pixels, are not as clear. The TV image also carries other information along the sides of the image, just as it carries time code information on the top and bottom. To keep things simple, you can say the image is 640 × 480.

Why does it matter? First, if you want the image to look good on the screen, you need to make sure everything matches. If you are going to create a graphic to use in your award-winning film, it has to be the right size; otherwise, it looks distorted. Trust me, I have made that mistake.

Note

Why is video 30 frames per second, not 24 as with film? The CRT's electron gun gets the electrons it shoots from the electricity in your house. The electric current alternates 60 times every second, which is why the field rate is 60 fields per second. Original black-and-white NTSC video had a frame rate of 30 fps. When color was introduced to the signal, however, older black-and-white TV sets could not display the images. Engineers found that color programs could be seen on those older sets if the frame rate was dropped to 29.97. Most editing software allows you to work with video at 29.97 fps, but make your life easier and just go with the 30 fps option. In Europe and many other parts of the world, electrical current alternates 50 times each second instead of 60. Because the flow of electricity is different, the field rate in those parts of the world is 50, so the frame rate is 25 fps. To address these differences, they need a different video standard. Two common ones in other parts of the world are PAL and SECAM.

Second, if you are going to compress your video for the Internet, keeping the same aspect ratio is helpful. If you shoot a video in one aspect ratio and deliver it in another, the images do not look good. I have made that mistake, too. For example, if you want to deliver video over the Internet, you need to make the video file small enough so that people can download it. One way to make the file smaller is to reduce the frame size (resolution) while maintaining the same aspect ratio. You can keep the same aspect aspect ratio by taking a video with a frame size of 640 × 480 and reducing the resolution to, say, 160 × 120. The video won't look distorted, but the file size will be small enough for viewers to download.

Just to make things a little more confusing, 4:3 is not the only aspect ratio in the television universe. HD uses 16:9 (see Figure 1-10). The 4:3 aspect ratio produces an almost square image; 16:9 is rectangular and is what you see if you watch a letterboxed video or DVD. Some digital video cameras shoot in 16:9, but that does not mean the video is HD. The last aspect ratio you need to know about is 3:2, which is the aspect ratio digital video uses (covered in more detail in "DV Format" later in this chapter).

FIGURE 1-10
HD uses a 16:9 aspect ratio

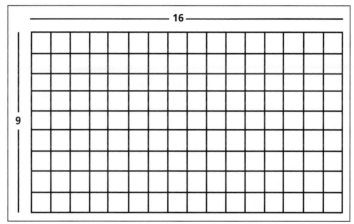

Analog Versus Digital Signals

What you hear and see moves through the air in sound waves or light waves. Until recently, recorded audio and video signals were basically just copies of those waves. This type of replication of those waves is called **analog** (see Figure 1-11).

FIGURE 1-11
An analog signal represented as a wave

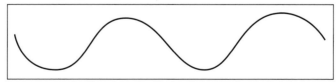

Although the analog signal is a faithful copy of the original signal, it has drawbacks. One drawback is that the signal might have little glitches, called **noise**, that are recorded, too. Noise is that annoying fuzz that looks like snow on the TV screen. When a signal with noise is put onto tape, the noise is also copied and continues to get worse in each copy (called "generational loss"). By the time you get to a copy of a copy of a copy, the picture and audio do not look or sound very clear. When you are editing analog audio or video, you are basically making copies of the originals.

Another problem with an analog signal is that it is **linear**, which means that to get to the end from the beginning, you have to go through the middle. You cannot just jump from the beginning to the end or go back and forth. An analog signal's linear nature can make it difficult to work with. One of my first assignments with a client was to edit a commercial that needed 25 different edits. The producer/director was busy talking on the phone while I worked. When the edit was finished, the producer/director realized he had given me the wrong tape for the second edit. The clip he wanted for the second edit was one second longer than the one he had given me. I had to start over because I could not just put the new clip in and be finished. I had to edit all the other pieces again that came after that edit.

The last problem with analog video is **bandwidth**. As explained earlier, bandwidth is like a pipe for information to travel through. If you want video to look better, you have to increase the amount of video information (data) in the video signal. If you increase the data rate (or amount of data moving down the pipe each second), you need to find a bigger pipe or find a way to reduce the data size so that it fits down the pipe. Analog video is difficult to make smaller, so you cannot make analog video look better and still have it fit in the same pipe.

Digital video solves these three problems with analog video. To see how, think of a video signal as a slanted line, as shown in Figure 1-12.

FIGURE 1-12
A section of an analog signal is like a slanted line

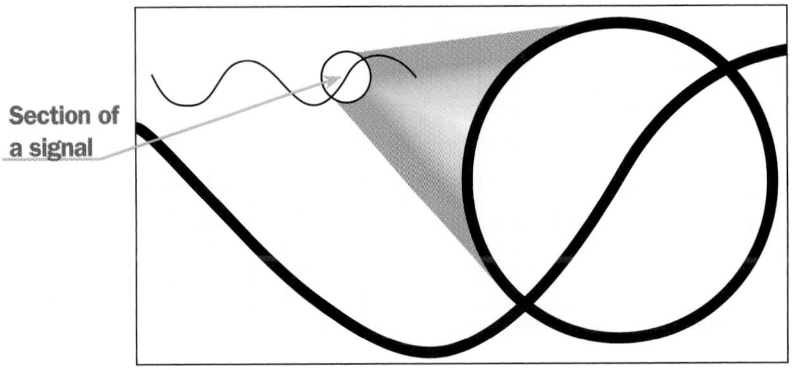

Section of
a signal

Color, light and dark areas, and even the noise mentioned earlier is on that slanted line. Analog video copies the line exactly as it is, including any noise or mistakes. Digital video uses a computer to pick out points along the line (see Figure 1-13) and assigns them a numeric value. This process is called **sampling**.

FIGURE 1-13
Sampling picks out points of the signal

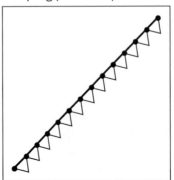

Digital video samples bits and pieces of the signal all the way down the slanted line. The number of points the computer picks along the line is called the **frequency**. The higher the frequency, the more points the computer picks along the line. As shown in Figure 1-14, a higher sampling frequency captures more information. Because digital video does not copy the line exactly, it does not pick up all the problems in the signal. That means noise is not copied from generation to generation.

FIGURE 1-14
On the left is a high sampling frequency, and on the right, a low sampling frequency

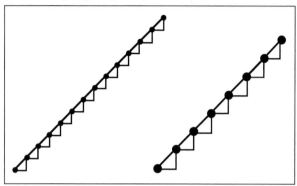

Digital video is also **nonlinear**, and digital editing systems are sometimes called nonlinear editing systems. Because the system is nonlinear, you do not have to go through the middle to get to the end. If I had been working on a nonlinear system for the project I talked about earlier, I could have just deleted the clip the producer/director did not like and inserted what he did want. Think of nonlinear editing like working with a word processor. With a word processor, you can cut a word or even a letter out of a sentence without having to retype the sentence. If you want to move an entire paragraph, all you have to do is copy and paste. It is the same with digital video. The computer does not care whether the data is video, audio, or text; it is all just random accessible data to be moved and worked with.

Digital video also solves the bandwidth problem. Digital video can be **compressed**, or made smaller, so more information (meaning a better picture and better audio) can be pushed through the same pipe analog video uses. Compression has a big impact on the quality of audio and video you watch on television and makes it possible to watch video over the Internet. You learn more about compression in Lesson 10, but for now, remember that compression makes it possible to have better video while using the same bandwidth.

Introduction to Digital Video

So far, you have learned about video basics and standards, such as scanning, frame rates, time code, aspect ratios, and so forth, but you have not learned about the DV format yet. Video comes in a number of formats or languages. Each format records the same images and sounds, but each does it in its own language. Video formats follow the same principles, such as 30 fps for NTSC video, but the way they follow these principles differs from format to format. For example, VHS and Beta are NTSC formats, but you cannot play a Beta tape in a VHS player because the two formats do not speak the same language.

DV Format

A few years ago, companies such as JVC, Sony, and Panasonic got together to decide on a new video format that uses digital technology. What they came up with is the DV format, which you should not confuse with the general principle of digital video. There are several digital video formats: DV, Digital Beta, and Digital 8, for example. The language that DV uses is different from the language that Digital Beta uses, even though both are based on the same digital video principles. Each digital format has its own language called a codec, which is a shortened form of "compression/decompression." A codec is basically software that tells the video format how to take the image apart during capture (that is, pack all the information in the image into a smaller "bag") and how to put it back together during playback. The codec determines the quality of video compression. Early codecs were not very good and produced blocky video, but newer codecs (discussed in more detail in Lesson 10) produce the high-quality video you see in digital cable, digital satellite, and DVDs.

Note

The DV format actually includes several formats, such as miniDV, DVCPRO, and DVCAM. These formats use the same codec but in a different way. DVCPRO, for example, uses a wider tape, which improves durability and editing performance. When DV is discussed in this book, it includes all those formats.

DV is compressed at a 4:1 ratio, or in simpler terms, the video is reduced to a quarter of its original size. Digital Beta, however, uses a compression ratio of 2:1, meaning the video is half the size it would be without compression. So which digital format is better? Digital Beta is supposed to be a much better format because it uses less compression, but most people cannot tell the difference.

The data rate for DV is 3.6 megabits (Mb) each second, or roughly four minutes of video per gigabyte (GB). The data rate affects not only image quality, but also "hard" costs. Because of the amount of compression for DV, you don't need as much hard drive space or specialized equipment. Lower hard cost means a huge cost savings when working with DV instead of other digital formats. DV is popular because of its high quality, but it is also much more affordable.

One way digital video formats compress video is through color sampling (also called "color space"). It is important to understand color sampling so that when salespeople start rattling off terms such as "as 4:1:1 color space," you know what they are talking about. The numbers refer to the amount of information used to describe the color in the video image.

To understand what this means, you need a little background in how people see things in the world around them. People see light and shadow as well as color. Light is referred to as luminance, and color is referred to in terms of hue and chrominance. Hue means the specific color, such as red, blue, or green, and chrominance means the amount of saturation. More chrominance means a deeper, richer red, for example; less chrominance means a lighter shade of red.

The human eye is more sensitive to light and darkness than it is to color. To see this for yourself, take a look at the color image in Figure 1-15.

FIGURE 1-15
A color image

Next, look at Figure 1-16, the same picture with all color information removed. As you can see, the human eye has no trouble seeing details in a black-and-white picture.

FIGURE 1-16
An image with the color information removed

If you look at the same image again with nothing but color information, however, making out details is difficult (see Figure 1-17).

FIGURE 1-17
An image with the light information removed

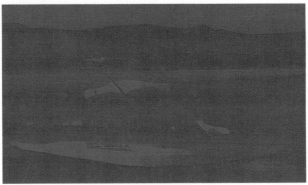

Digital video takes advantage of the way people see by storing all the light and dark information (luminance) and keeping only part of the color information (chrominance and hue). Going back to the 4:1:1 color space, the first number represents luminance, the second number represents chrominance, and the third number represents hue. So a 4:1:1 color space means that digital video keeps all the luminance information and just a quarter of the chrominance and hue information. Digital video gets rid of information that most people's eyes are not sensitive to. By comparison, some expensive digital video formats have a 4:4:4 color space, which means they do not get rid of any color information.

One final point about the DV format: The aspect ratio is 3:2 with a resolution of 720 × 480. This means DV is similar in height to standard definition video but a little wider (by about 40 pixels.) The aspect ratio is not a big deal until you are trying to create graphics or use your editing system. Some editing systems do not allow you to edit 4:3 video with 3:2 video; others do, but you have to go through a process called rendering first. It does not make much difference when you are watching your video on television, but knowing the aspect ratio can save you a lot of trouble when you are trying to figure out why your graphics do not look right or why your editing system is making you render everything.

> **Note**
>
> Just because manufacturers got together to create a standard for DV does not mean they did not go home and create their own variations of the standard. As mentioned earlier, DV comes in many different versions, such as miniDV, DVCAM, and DVCPRO. This book focuses on miniDV because it is the least expensive, and by far the most accessible, digital video and DV format.

FireWire

One of the major advantages of DV over most other digital video formats is how you get it into your computer. To get analog and some digital formats into a computer, you need a capture card. A **capture card** converts the video into digital information the computer can work with. Some capture cards are expensive and rely on specially built and formatted computers. DV, however, uses a **FireWire** capture card. Apple (you know, the computer company) developed FireWire for high-speed data transfer, but few people took notice at first. Different providers will refer to FireWire differently, for example, an organizational standard is IEEE 1394 or 1394 and Sony refers to it as iLink. Eventually people realized it would be an easy, inexpensive way to integrate video into personal computing. DV latched on to FireWire, eliminating the need for expensive capture cards. Being able to capture video through a FireWire port, which is standard on many computers, is one of the great advantages of DV. The video comes in as a digital signal and stays that way.

SUMMARY

In this lesson, you learned:

■ A television contains a cathode ray rube (CRT), which includes a screen and an electron gun. The electron gun shoots an electron beam that lights up phosphorus pixels on the screen. The pixels are organized in scan lines across the screen. The odd lines are scanned first, followed by the even lines.

■ One interlaced video frame consists of two fields. The first field is created when the odd-numbered lines have been scanned. The second field is created when the even-numbered lines have been scanned. Video is displayed at 30 fps, or 60 fields per second.

■ A progressive scanned frame of video is scanned in a single pass, with each line being scanned in order. Progressive scanned video can be displayed at 24, 30, or 60 fps.

■ Analog video makes a faithful copy of the light waves you see. Digital video picks points along those waves to capture. The sampling frequency determines the number of points captured along the wave.

■ Analog video is linear, meaning you have to go through the middle to get to the end. Digital video is nonlinear, meaning you can access any point at any time.

■ Digital video can be compressed, or made smaller, but analog video can0=not.

■ DV is a format of digital video that is different from other digital video formats. DV captures the video at 3.6 Mb, uses 4:1:1 color space, and is displayed in a 3:2 aspect ratio with a display resolution of 720 × 480.

■ The DV format includes miniDV, DVCPRO, and DVCAM.

■ DV connects to editing stations through FireWire. FireWire eliminates the need for expensive capture hardware.

VOCABULARY *Review*

Define the following terms:

Analog	Frame	Noise
Bandwidth	Frame accurate editing	Nonlinear video
Capture card	Frequency	Persistence of vision
Cathode ray tube (CRT)	High definition (HD)	Pixel
Codec	Hue	Progressive scanning
Color sampling	Interlaced video	Red, green, blue (RGB)
Color space	Linear video	Resolution
Compressed	Luminance	Sampling
Field	National Television Systems	Scan lines
FireWire	Committee (NTSC)	Time code

REVIEW *Questions*

MULTIPLE CHOICE

Select the best response for the following statements.

1. The aspect ratio for DV is _____.
   A. 4:3
   B. 16:9
   C. 3:2
   D. 4:4

2. DV is the name for which of the following?
   A. any type of video
   B. a specific digital video format
   C. any type of digital video
   D. only miniDV

3. FireWire is used for which of the following?
   A. connecting video equipment to your computer
   B. connecting video equipment to a capture card
   C. connecting video equipment to a television

4. Color sampling or color space refers to which of the following?
   A. the color in the space around the subject
   B. how color looks in the camera viewfinder
   C. a way to color correct video
   D. the way a format captures color information

TRUE/FALSE

Circle T if the statement is true or F if the statement is false.

T   F   1. NTSC video is displayed at 60 fields per second.

T   F   2. The frame rate for NTSC video is the same as the frame rate for film.

T   F   3. The frame resolution for DV is 720 × 480.

T   F   4. All video shot at 16:9 is HD video.

T   F   5. DV is an HD format.

T   F   6. Digital video uses sampling to capture video information.

T   F   7. Analog video has a different frame rate from digital video.

T   F   8. Frame accurate editing is possible because of a time code.

WRITTEN QUESTIONS

Write a brief answer to the following questions.

1. Briefly describe scanning.

2. Briefly explain what makes DV different from other digital formats.

3. Describe the differences between analog and digital video.

4. Describe the process of sampling.

PROJECTS

Go to an electronics store and watch TV. Take a look at traditional tube TVs, and then look at an HD TV. What are the differences? What is the same? Which one looks better? Can you tell the difference between a 720p image and a 1080i image? Which do you prefer? If you can, go to a movie at a theater. Can you tell the difference between the images on the screen and what you looked at on a TV screen?

TELLING A STORY

LESSON 2

OBJECTIVES

Upon completion of this lesson, you should be able to:

- Identify the basic elements of a story
- Write a script for a news broadcast in Final Draft AV
- Write a script for a short documentary in Final Draft AV
- Develop a narrative fiction story
- Write a script for a short narrative fiction video in Final Draft

Estimated Time: 1.5 hours

VOCABULARY

Active sentence

Antagonist

Conflict

Content

Documentary

External conflict

Internal conflict

Narrative

Passive sentence

Protagonist

Script

Story

Style

Introduction to Story

So far you have learned about video and DV and why DV is such a great technology. When you finish all the lessons in this book, you will have the skills to produce a video with DV, whether it is a news broadcast, a sportscast, a documentary, or a narrative video.

But where do you start? After you have decided what kind of video you are going to produce, you need a plan. For a video, this plan is a script. A script helps organize ideas, identifies what the video will look and sound like, and makes sure everything is connected and flows in a logical order. Writing and altering a script is much less expensive than re-shooting and re-editing a video. A script also communicates your ideas to people you might be working with, such as camera operators and actors. People make videos without scripts all the time, but they are usually family videos that only family members watch. If you want a larger audience, you need a script.

In this lesson, you concentrate on writing scripts for news broadcasts, documentaries, and narrative (fictional) videos. Before examining the different types of scripts, it is important to understand what all three types of productions have in common: they all contain a story.

Elements of a Story

DV and other types of video are tools used to tell a story. You might watch family videos a hundred times and enjoy them, but you do not see many family videos in theaters or on television. If you want other people to watch your news broadcast, documentary, or narrative, you need to tell a story.

But what is a story? People tell and hear stories all the time. Sitting around during lunch and talking about what you did last night with friends is telling a story. A story is simply a retelling of events that happened. Even retelling unimportant events can be considered a story. Sometimes storytellers are so good, that even unimportant stories, such as how they picked their doughnut that morning, make the audience sit on the edge of their seats. Somehow a good storyteller is able to make the problem of picking a glazed doughnut instead of a cake doughnut an exciting story. How does a storyteller do it? What does the storyteller know that helps her or him tell a good story?

First, a storyteller knows that any story is about **conflict**. If you get up and walk across the room to pick up a pencil and then return to your desk, is that a story? If a reporter stands at the end of the runway and reports on every flight that lands safely, is that a news story? The answer to both questions is no. How interesting is it to watch somebody walk back and forth across the room picking up pencils? A conflict describes when and how things go wrong.

Second, a storyteller knows how to make you care about the story. A lot of movies have great special effects and an amazing soundtrack, but if you do not care about the characters, you do not care about the story, even if it is about somebody saving the world. The best news stories make you think about how the story affects you, and then you care about what is going on. Stories are about characters facing conflicts and having to make choices about the situation they are in. Even if you have not been in the same situation, you can understand the choices the characters have to make. Note that I said characters, not people. Some of the best movies (especially family-oriented movies) are about objects or animals that are given human qualities to which you can relate.

Lastly, a storyteller gets to the point. Have you ever watched a movie and were not sure what it was about when it was over? The movie is so packed with special effects, beautiful people, double and triple crosses, and snappy dialogue that you do not know what is going on. A storyteller does not waste your time telling you anything but the story.

STEP-BY-STEP 2.1

One of the best ways you can improve your storytelling skills with video is to watch what others have done and identify what they do right and wrong. You do not have to watch award-winning videos either. Some of the best movies to learn from are the worst movies ever made. Sit through a terrible movie and figure out what you could do to make it "work."

1. Open a new document in your word processing program.

2. Watch a movie, TV show, or newscast, and see whether the storyteller includes the three elements of a story.

3. Notice whether you can identify the conflict, whether you care about the story and why, and whether time is spent on information that has nothing to do with the story. Write what you have identified in a word processing document.

4. If the video does not include these three elements of a story, explain how the storyteller could have improved the story. Write your explanation in a word processing document.

5. Save the document as **SBS2-1** and close the program.

Types of Productions

In this lesson, you learn about three types of production in which scripts are written: news broadcast, documentary, and fictional narrative. Each of these types of productions are described in further detailed in the following sections.

News

A newscast has to be scripted to run smoothly. The director, the tape operator, the camera operator, and everyone else on the set have to know what the plan is and what is coming next to make things work. They have to know which anchor is taking which story, when the weather reporter is coming on, and when to go to highlights during sports and which highlights to roll. Everything has to be timed to the second to make sure the newscast goes as smoothly as possible.

Television news has followed the same format for years: a news anchor or talking head telling or introducing the news story with video clips of the actual event or interviews to illustrate the story. For example, you should not just talk about a flood, but you should also show the houses floating downstream. You want the audience to experience the story, not just hear about it. The guideline is "Do not tell me. Show me."

This section is divided into two parts: content and style. **Content** covers *what* you write; **style** covers *how* you write it.

Content

I cannot tell you what your news story should be about. However, I can tell you what you need to include in the story: what, who, where, when, how, and why. The audience has to know what happened, who it happened to (or who did it), where it happened, when it happened, how it happened, and why it happened. You probably have heard this list before. Think of it as questions you need to answer for each story.

You will not always have all that information, but make sure you give the audience as much as possible when you write your story. Take a look at two ways to tell the same story.

"There was an accident on Highway Six today."

What information is included? The only thing you know for sure is that there was an accident. You have a general idea of where and when, but not much more. What kind of accident was it? Did somebody stub his or her toe? Was anybody hurt? You do not know. Now try it this way:

"A truck jackknifed on Highway Six three miles east of Dodge early this morning. State police believe the driver fell asleep and slammed on his brakes to avoid hitting another car. The driver was not wearing a seatbelt. State police are withholding the name of the truck driver until his company has been notified."

What information do you have in this version? You know what kind of accident it was, you know where the accident happened, and you have a better idea of when, why, and how it happened. You still do not know who it happened to, but you know why you do not have that information.

Style

Remember, the guidelines in this section are the basics. Each news station has its own style guide, so make sure you have read and understand what the station wants.

The six points are:

- Write in a conversational style
- Use active sentences
- Start strong
- Write simply
- Do not be wordy
- Write to be spoken

Let's look at each one individually.

Write in a conversational style

The audience *listens* to the news, so they need to understand it the first time they hear it.

"Because he had a headache, the president of France went to bed early."

Would you talk to a friend like that? No. How would you say it?

"The president of France went to bed early because he had a headache."

Writing conversationally does not mean, however, that you should present your news story like this:

"Dudes, like, last night, man, this dude, the president of France or something, had, like, this nasty, I mean really nasty, headache. He had to hit the sack early."

The best way to write conversationally is to listen to what you are writing. Read it out loud and make sure it is clear the first time you hear it. This advice is not unique to writing newscast scripts; it applies to all writing. Sound professional and conversational, but do not use long sentences and big words to try to sound sophisticated.

Use active sentence

An **active sentence** follows the basic formula: subject, verb, object (S-V-O). Take a look at this sentence:

"The ball was played with by the cat."

That sentence is not an active sentence but it is a **passive sentence**. The object (the ball) is first, followed by the verb (played), and the subject (the cat) is at the end. How do you make the sentence active? Write it in subject, verb, object order:

"The cat played with the ball."

Passive sentences are longer, harder to read, and more difficult to understand when heard. Keep your writing active. Sometimes it is okay to use a passive sentence, but not very often. For example:

"Suzie Martin was found early this morning."

The sentence is passive, but the emphasis is on Suzie Martin. If Suzie had been missing for three days, everybody would pay close attention at the mention of her name. You could rewrite the sentence to make it active, but it might lose some of its impact and audience:

"Rescuers found Suzie Martin early this morning."

Another instance when a passive sentence would be acceptable is if you do not know who did what. For example, the sentence might read:

"Suzie Martin was found early this morning near Mount Saint Helens."

It might not be clear who found Suzie, and saying "Somebody found Suzie Martin" would only create more questions, such as "Who found her?" Usually, the best way to approach any writing is to start with the S-V-O formula and make changes from there.

STEP-BY-STEP 2.2

In this Step-by-Step you change passive sentences into active sentences.

1. Open a new document in a word processing program of your choice.

2. Change the following five sentences from the passive voice to the active voice.
   a. The fire was put out in a matter of minutes by the fire department.
   b. In order to get a good score, John cheated on his test.
   c. The garbage can was run over by that blue and green car.
   d. The protest group was surrounded by the police and forced to leave the capital building.
   e. Police were surprised by Mac Strange, the notorious bank robber.

3. Save the file as **SBS2-2** and close the program.

Start strong

Start strong to catch the audience's attention and fill in the details later. You have only a couple of seconds to catch the audience's attention. You do not want to waste the first sentence on a weak introduction. Again, this rule does not apply only to newscasts, but to all writing. The first example has a weak introduction.

"Searchers near the Red Canyon saw a girl walking along the river early this morning. The girl was Suzie Martin, who had been missing since Thursday."

The second example is much stronger because it gets right to the point of the story.

"Suzie Martin was found early this morning near the Red Canyon. Searchers found Suzie, who had been missing since Thursday, walking along the river."

People would pay close attention to the story if they heard the name of the girl who had been missing right off the bat. It takes a while for the first example to get to the point.

Write simply

Use big words such as "implemented" or "endeavored" when you are trying to impress your boss, but leave the thesaurus on the shelf when you are writing for a newscast. This rule goes along with using a conversational style. Which one works better?

"The president asked Congress for another five billion dollars for education."

"The president implored Congress for an additional five billion dollars for education."

The second sentence may be harder for people to understand because of the word "implored". Not everybody knows what it means. You will lose your audience if they have to run out to find their dictionary.

Do not be wordy

Get to the point and stick to it. Tell the story as simply as possible, and stay away from trivia and little bits of information that do not have anything to do with the story.

"Three people were seriously injured in a single car accident on Route Seven last night. Route Seven goes through the hometown of our own weatherman, Jacob Smart. Witnesses say the car was traveling at about ninety miles an hour when it lost control and rolled over."

The sentence about the weatherman has nothing to do with the three people who were serious injured and trivializes their injuries. Unless the information has direct bearing on the story, do not throw it in.

Write to be spoken

News scripts are written to be read out loud, usually from a teleprompter, so make it easy to say what is written. For example, 1600 can be read as "one thousand, six hundred" or "sixteen hundred," so write out numbers to make sure the reader knows exactly what to say. Include pronunciation guides or other information to make sure names are pronounced correctly. For example, you might want to spell out the name Andre as "On-dray," just to make sure it is pronounced right. The audience grades you on your pronunciation, not your spelling.

Writing is an art that has to be studied and practiced. Do not think that just following these six steps will win you any awards. Good writing takes years of practice, but you can do it if you put in the effort. Another factor to consider in writing a good news story is to distinguish between news and opinion.

News vs. Opinion

Look at these two approaches to the same story:

"Mayor Dunn has been accused of misappropriation of city funds in the past, but no charges have been filed."

"This is the third time Mayor Dunn has been accused of stealing taxpayers' money. Although she has avoided prosecution in the past, it is only a matter of time before investigators find enough evidence to convict her."

Which approach is neutral?

Separating personal beliefs from how news is reported is often difficult. Recently, both sides of the political fence have accused the media of supporting the other side's cause. Some TV and radio news programs try to manipulate listeners into believing a certain point of view, but people usually prefer to hear the facts and decide what they believe themselves.

There is a difference between a reporter reporting the news and a commentator trying to shape opinions; this point of view is decided when the script is written. Keep in mind that your audience can discern the difference. If you do not like the mayor, the audience will pick up on that. You could lose your audience if they do not agree with your political point of view. You might not care if you lose your audience, but then again, if you do not have an audience, you might not have a job either. It is easy for news to go from being reporting of the facts to being propaganda. Make sure you know what you want to accomplish and write your script accordingly. Be as fair and balanced as you can if you want to be a reporter. There is nothing wrong with being a commentator, but make sure you (and your viewers) know the difference between news and opinion.

Writing a Script for a News Broadcast

How does what was just covered relate to the three things a storyteller knows? First, when reporters go after a story they have to ask themselves whether they have a story. Is there conflict? Is something out of the ordinary going on? If the best an anchor can report is "Two hundred cars drove by my house and not one crashed," he would lose his job. A good reporter is able to communicate the conflict clearly to the audience.

Second, a reporter has to know his or her audience and how the story applies to that audience. If a reporter in Los Angeles tells the story of a water shortage in Texas, he or she better be able to help his or her Los Angeles audience understand how the shortage in Texas affects them in Los Angeles.

Third, news reporters, at least good ones, jump right to the heart of the story. They do not waste your time telling you about how they did their makeup that morning.

STEP-BY-STEP 2.3

Let's practice. The following are the facts of a recent news story.

1. Use the following information to write a script for a news anchor. Create the script in a word processing program of your choice.

Fifteen-year old Adam Taylor was fishing with his brother and father early this morning near Lake Water, elevation 6463 feet above sea level. Adam and his seventeen-year old brother, Tyler, got into an argument over whether to use real or artificial worms.

"Adam is a purist, you know," said Tyler. "He has to spend hours digging for worms and such. He has to have real worms, or the experience absolutely is ruined for him."

Adam threw water at his brother and stormed away from the others, going southwest around the lake. Adam was wearing jeans and a white T-shirt with the words "Vote for Pedro" in fuzzy red letters.

The father and brother continued fishing for about an hour before they realized Adam had not returned to fish with them.

"I thought he was pouting in the bushes or something," said Adam's father. "He likes to sit in the bushes and pout. He is really good at it, too. Why once, when he was six, he pouted in the bathroom for three hours before we changed the channel."

Adam's father and brother stopped fishing, walked to the car and put their gear away, and then walked in the direction Adam had gone.

"We argued for a while before we looked. I thought he went that way," said Tyler, "but Dad said he went the other way. So we went the other way."

After two hours of searching for Adam, the father and brother stopped looking and called search and rescue.

Currently, 113 searchers, including three women from Siberia, are combing the area around Lake Water looking for Adam. The major concern at this time is a cold front moving in from the Northwest. The weather forecast calls for snow tomorrow and overnight lows below freezing.

2. Save your work as **SBS2-3** and close the program.

Did you simplify? Did you write out numbers? Did you start strong? Go over the six points again and compare what you did. If you can, have somebody else go over what you wrote while you go over what they wrote. Give each other positive feedback while making sure you both know how you can improve. Remember, you are both supposed to be learning. Never make fun of anybody for trying to learn something new.

Documentary

The easiest way to describe a documentary is to say it is nonfiction film or video. Like the news, documentaries are based on facts. The facts in a documentary, however, can be recent or historical events. Documentaries also cover subjects in more depth than the evening news does. Documentaries are a good opportunity to examine a subject more thoroughly.

Some people think a documentary is just a lot of video footage put together without a script. Even though you are using interviews and facts, you still need a script. I have a couple of friends who are successful documentary filmmakers, and one of their most common complaints is that they have trouble getting the script right. For example, I decide my grandmother has lived an interesting life and is worthy of a documentary. She has a trunk full of home movies I can use, and I can set up my camera and ask her a few questions. Then I can take all that footage and start editing. It sounds simple, right? That approach might take a long time to finish, however, because I do not have a clear goal in mind.

A better approach is to write down what I know about my grandmother. Instead of trying to make a documentary about her entire life, I should focus on a specific timeframe. I would do that before watching any of her home movies. My grandmother lived in a cabin in the Yukon with my grandfather and their four young boys, so a documentary about those experiences could be interesting. I could contrast how she lives now and show how that early time period formed and shaped her life.

Every writer has his or her own way of developing a story, but I have found the following process is helpful in developing a story. You do not have to follow these steps exactly, but you can use them to develop your own method.

1. Write down what the documentary is going to be about. Identify what you want the story to say. For my documentary about my grandmother, my focus would be how her experiences in the Yukon affected the rest of her life.

2. Research the topic and put the facts on paper. I would find out when and why my grandparents moved to the Yukon. I would find out what they did when they were there and what life was like for others living in the area at the same time. Did they have running water? Did they have a grocery store nearby? Did they have any neighbors? At this point, I also would watch her home movies and go through her photo albums to find which visual elements could be used to tell the story.

3. Write questions to ask your subject and any others who have pertinent information to be included in the documentary. I would ask my grandmother, father, and uncles about their experiences and what my grandmother was like before and after they lived in the middle of nowhere. The script is more of an outline at this point, but it is enough to begin shooting.

> **Hot Tip**
>
> You might start off with one idea, but when you gather the facts and footage, you might realize your initial assumptions were wrong or you have found a more interesting story to tell. Be prepared to change your focus if necessary.

4. After you have interviewed your subjects and acquired all the footage and material you want to include in the documentary, you should then revisit your script. Determine whether your story is still what you thought it was going to be. If you can tell the story you originally set out to tell, move on to the next steps. If you have to change your approach, go back through the earlier steps and clarify your story. You cannot tell a story if you do not know what it is about, and you will end up spending a lot of time reviewing footage without making any progress.

5. Organize your footage. Write down what you have and how it will be put together. This step helps you locate what you want, spot any problems in the story, and identify any missing visual elements. You might find that you need more footage or information.

At this point, you have a rough cut and might decide you want a voiceover (narration) to pull everything together. If so, write it, remembering the six guidelines for writing news scripts covered previously.

Propaganda or Information?

I recently watched a documentary about a controversial government project. The documentary filmmakers appeared to present fair and balanced information, but clearly they were strongly opposed to the project. They intended to convince the audience that the project was a bad thing by giving little time for those in favor of the project to express their opinions by showing the positive emotions of the opposing people but only the negative emotions of the people in favor of the project, by letting the opposing side have the last word, and by portraying the negative images of the people in favor of the project. Before I watched the show, I had no opinion about the project. However, after watching the show, instead of being opposed to the project because of the information presented, I became angry. I felt they were trying to manipulate me. As with the news, there is a fine line between presenting information and creating propaganda. If you want to create propaganda, be aware that the audience can tell what you are doing.

Writing a Script for a Short Documentary in Final Draft AV

News and documentary scripts are discussed together in this lesson because they share the same script format. You could use any format you want, but the two-column format shown in Figure 2-1 is the most common.

FIGURE 2-1
Two column format

Agency		Writer	
Client		Producer	
Project		Director	
Title		Art Director	
Subject		Medium	
Job #		Contact	
Code #		Draft	

VIDEO	AUDIO

The two-column format is simple yet effective. Write the visuals (what people will see) in the column on the left. Write what will be said in the column on the right. Add rows to the columns to create cells that align the visuals with the audio. See Figure 2-2. For example, if you show a horse but are talking about a cat, you will confuse the audience.

FIGURE 2-2
Add rows to the columns that align the visuals with audio

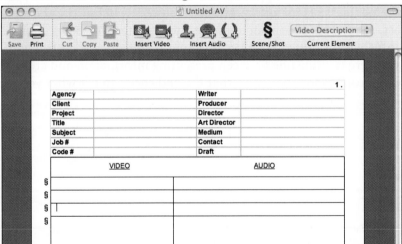

When I first started writing scripts, I had to format them myself, making sure every line was where it needed to be. I'm glad I had a word processor as I would have given up scriptwriting if I had to do it on a typewriter. Now, software takes care of the formatting for you. The software I use is Final Draft and Final Draft AV. Final Draft is for Hollywood-style single column scripting (you will see the difference when we cover Fictional Narrative scripting), and Final Draft AV is for two-column scripting. You will start with Final Draft AV. First, take a quick look at the Final Draft AV layout in Step-by-Step 2.4, which is similar in Windows and Macintosh.

STEP-BY-STEP 2.4

In this Step-by-Step activity, you open Final Draft AV 2.5 and a new document. Instructions for opening Final Draft AV and a new document are provided for both Macintosh and Windows XP. Follow the instructions for your operating system.

Macintosh instructions

Follow these steps to start Final Draft AV in Macintosh. (If Final Draft AV Setup Assistant window appears, follow the on–screen instructions.)

1. To start Final Draft AV, double-click the **Macintosh HD** icon. (If the hard drive name has been changed, find and double-click it.)

2. Click or double-click the **Applications** icon in the frame on the left, or click or double-click the **Applications** folder in the second frame from the left.

3. Locate and click or double-click the **Final Draft AV** folder.

4. Click or double-click the **Final Draft AV** icon.

5. Final Draft should open a new script automatically when you start the program, but you also can use any of the following methods to open a new script:
   a. Click **File**, then click **New** from the menubar.
   b. Press **Command+N**.

> ### Hot Tip
>
> You can place an alias (or shortcut, for those used to Windows terminology) on the Dock by dragging the icon onto the Dock. You can also click the icon, and then press Command+L to create an icon that you can drag to the desktop. To start the program, just double-click the icon on the desktop or Dock.

Windows XP instructions

Follow these steps to start Final Draft AV in Windows XP:

1. Click **Start** and point to **All Programs**.

2. Click the **Final Draft AV 2.5** icon.

3. If you are starting the program for the first time, a number of dialog boxes appear. If an End User License Agreement dialog box appears, click the Accept button. Next, a Welcome to Final Draft AV dialog box appears, click the **Demo Mode** button, then a Welcome dialog box appears, click **Continue**.

STEP-BY-STEP 2.4 Continued

4. Final Draft AV should open a new script automatically when you start the program, but you also can use any of the following methods to open a new script:

    a. Click the **New** icon at the left of the toolbar. (It looks like a piece of paper with the top-right corner folded down.)

    b. Click **File**, then click **New** from the menu bar.

    c. Press Ctrl+N.

> **Hot Tip**
>
> You can create a shortcut for Final Draft AV by following Step 1, and then right-clicking the Final Draft AV 2.5 icon. On the shortcut menu, click Create Shortcut. Click and drag the new shortcut to the desktop.

5. Leave the program open for the next Step-by-Step activity.

Now that you have a new script ready to go, let's start writing a script.

STEP-BY-STEP 2.5

In this Step-by-Step, you write a script using Final Draft AV. The following steps work with both Windows XP and Macintosh.

1. With Final Draft AV open and a new document on the screen, the cursor (called the "insertion point" in Final Draft AV) should appear in the VIDEO column. If it is not there, click inside that column.

2. Type **Grandma's house from the front yard.** Press **Tab** to move the cursor to the AUDIO column. Notice that a colon (:) is inserted automatically.

3. Type **Narrator.** Final Draft underlines the text automatically and adds a colon at the end of the word "Narrator." Press **Tab** to move the cursor to the next line.

4. Type **Grandma moved to this house in 1998 after Grandpa died.** Press **Tab** to move the cursor to the VIDEO column.

5. Type **Montage of Grandma's house: the bathroom, the kitchen, her television.** Then press **Tab** to move to the AUDIO column.

6. Type **Narrator** and press **Tab**.

7. Type **This house has all the modern conveniences: electricity, central air, a modern kitchen, a 56-inch screen television, and satellite television.** Press **Tab** to create a new scene, in which you will create a short conversation between two characters.

8. Type **Still image of house in the Yukon**, and press **Tab** to move to the AUDIO column.

STEP-BY-STEP 2.5 Continued

9. Type **Grandma.** Click the **Insert Parenthetical** icon on the toolbar or use **Command+4** (MAC) or **Control+4** (Windows). You should see two parentheses with the cursor between them as shown in Figure 2-3.

FIGURE 2-3
Two parentheses

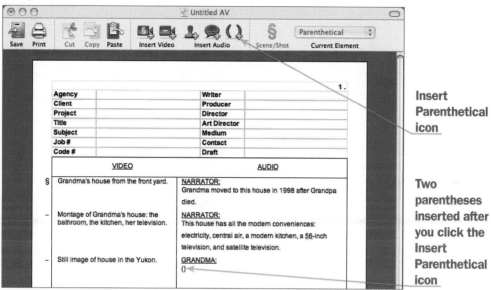

10. Type **V.O.**, which stands for "voiceover." Grandma will be saying something, but the audience will not see her yet. Press **Return** or **Enter**. You are now ready to write your dialogue.

11. Type **We lived in the Yukon for seven years.** and then press **Return** or **Enter**. This moves the cursor down to the next line and gets you ready to introduce a new voice.

12. Type **Me.** Now you are in the video. Click the **Insert Parenthetical** icon on the toolbar or use **Command+4** (MAC) or **Control+4**.

13. Type **O.S.** which stands for off screen and press **Return** or **Enter**.

14. Type **Why did you move up there?** and press **Return** or **Enter**.

15. Type **Grandma** and Click the **Insert Parenthetical** icon on the toolbar or use **Command+4** (MAC) or **Control+4**.

16. Type **O.S.** and press **Return** or **Enter**.

STEP-BY-STEP 2.5 Continued

17. Type **That was your grandfather's idea.** Press **Return** or **Enter**. Your document should look like Figure 2-4.

FIGURE 2-4
Example of documentary script

Agency	Writer
Client	Producer
Project	Director
Title	Art Director
Subject	Medium
Job #	Contact
Code #	Draft

VIDEO	AUDIO
§ Grandma's house from the front yard.	NARRATOR: Grandma moved to this house in 1998 after Grandpa died.
— Montage of Grandma's house: the bathroom, the kitchen, her television.	NARRATOR: This house has all the modern conveniences: electricity, central air, a modern kitchen, a 56-inch television, and satellite television.
— Still image of house in the Yukon.	GRANDMA: (V.O.) We lived in the Yukon for seven years. ME: (O.S.) Why did you move up there? GRANDMA: (O.S.) That was your grandfather's idea.

18. Click **File**, then click **Save** from the menubar.

19. Type **SBS2-5** as the name of the file and check with your instructor to find out where to save the file.

20. Click **Save**. Close the document and program.

One more thing before we move on to the next section. How does a two-column news script differ from a two-column documentary script? Look at Figure 2-5 for examples and see if you can tell any differences.

FIGURE 2-5
Example of two-column documentary and news script

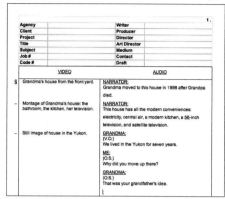

Fictional Narrative

The story structure of most Hollywood movies is the classic structure, with a clear beginning, middle, and end. The story is typically easy to follow, and you know who the heroes and villains are. Often the story involves some sort of quest, with the villains trying to stop the heroes from achieving their quest. Although this basic form of storytelling sounds simple, there is an art to doing it well.

Long before movies and plays, there were storytellers, such as Homer. The classic structure was developed centuries ago and is still around because audiences relate to it, understand it, and expect it. A story can be told using any number of different structures, but the farther you stray from the classic structure, the smaller the audience becomes. This lesson covers the classic story structure as a starting point.

Beginning - Protagonist and Antagonist

The classic story begins by introducing the main characters and the conflict. The main character is known as the protagonist. At the beginning of the story, the audience needs to have an understanding of what is normal and acceptable in the protagonist's life. For example, the protagonist Suzie has an office with a box of pencils on the other side of the room. She can get up and get a pencil from the box any time she wants. After the audience gets an idea of what Suzie's life is like, the beginning is over, and it is time to introduce the conflict and move into the middle of the story.

The middle starts when the protagonist's world or situation changes. What if Suzie walks over to get a pencil, but suddenly, Mr. X, the bad guy (known as the antagonist) stops her? He tells her she will never be able to get a pencil again because he will stop her if she tries.

You have a conflict when Suzie decides that no matter what happens, she is going to walk across the room and get another pencil. If she decides not to get any more pencils, you do not see the bad guy again, and you do not have a story. Do you have a story if she decides she is going to get another pencil? You do as long as Mr. X comes back. You have heard the saying "it takes two to tango." In other words, to have a conflict, there must be opposing forces. Because of this conflict, Suzie's world has changed, and she will do anything to put her world right again.

Conflict

Most movies have an **external conflict**, which is something or someone outside the protagonist that prevents them from obtaining their goal. That something or someone can be a bad guy, a huge storm, a monster, etc. Action films usually rely on an external conflict. Some films have an **internal conflict**, which is something within the characters themselves that keeps them from obtaining their goal, such as fear, a lack of confidence, or simply being tired. External conflicts are easier to dramatize and show, but the most effective stories are usually a combination of external and internal conflicts, such as a detective who has to prove someone is responsible for a crime but is afraid of being wrong because he put an innocent man in jail once.

The Middle

The middle of the story is the buildup of tension between the two opposing forces. It typically is the weakest part of the story. Suzie wants a pencil and continually finds new ways to get one, but the bad buy finds new ways to prevent her from getting it. It is important that the audience believes Suzie will achieve her goal. If the audience does not believe in her, what is the point? The odds against her might be so huge that the audience knows deep down she will never succeed, but they have to at least have some hope. The audience also has to believe that the antagonist could actually prevent her from achieving her goal. Mr. X might be bigger, stronger, and smarter, for example, but Suzie might have more willpower.

The End - Climax

The middle ends when the characters come to a point where there is only one option and only one thing left to do. At this point, the story reaches its climax, and only one can win. It is the final attempt, and either Suzie gets what she wants or Mr. X gets what he wants. If Suzie succeeds, you have a positive, or happy ending. If Mr. X wins, the story has a sad, or negative ending.

A good storyteller knows that when the audience cares about the characters it is because they relate to them and want them to succeed. You might not ever have Suzie's problem, but you can sympathize with her plight because you can envision her struggle as your struggle. If Suzie wins, you cheer because you feel as though *you* have won.

People also care about characters because of the choices they make. If, for example, Suzie chooses to beat up Mr. X the first time she sees him, you think she is tough. If she chooses to sink back to her desk without doing or saying anything, you label her as submissive. It is important to be consistent with the choices characters make. If Suzy is submissive to Mr. X in one scene, she cannot be tough in the next scene without a reason that makes sense. People expect and want characters to change, but they have to believe the character can change. Remember: Actions speak louder than words. In fictional stories, you have to "see" the change.

A fictional narrative is based on ideas that have to seem as realistic and powerful as nonfiction. If the fictional world you create does not ring true to the audience, then it is not accepted as reality. So how do you start? These steps are just one way to approach a story. Every writer approaches storytelling differently, so feel free to go with what works for you. Make sure you put your story on paper (or in a Word document) to help you remember your ideas and develop them further.

S TEP-BY-STEP 2.6

1. Open a word processing program and a new document.

2. First, give the story a setting. It helps to know when and where things are happening. A story about a woman in a small town in Iowa is going to be different from a story about a woman in New York City, and a woman's attitude toward the world today is different than it was 200 years ago. The conflict might be universal, but how the character deals with the conflict will vary because of the setting.

3. Give the character a name and an identity. You can rename the character later, if you like, but this step helps you identify with your character.

4. Specify in detail what the character wants. If you decide the character wants to be happy, explain what you think will make them happy, such as, "Suzie wants to be happy by getting a new pencil every day."

5. Identify who or what is preventing the character from reaching that goal. If the antagonist is a person, give the person a name.

6. Develop a logical progression of how the protagonist will overcome the antagonist. For example, Suzie tries to sneak a pencil, but that does not work, so she tries to do it at night. That does not work, so she brings a dog to keep Mr. X out. The progression should be logical for the character. If Suzie is a shy, quiet person, she probably would not begin by being an aggressive person. She might get to that point eventually, but her actions have to make sense for her character.

7. Know how the story will end. Often the story leads to what you think will be a logical, satisfying ending, but you end up with something that makes you wonder why you wasted your time. Make sure the ending makes sense and follows the rest of the story.

8. Save your story as **SBS2-6** and close the program.

Writing a Script for a Short Narrative in Final Draft

So far you have learned about the development of a narrative story, but not about the actual writing process. You can use the two-column script format you learned about earlier, but it is not the "official" narrative script format. The official format looks like Figure 2-6

FIGURE 2-6
Narrative script format

```
INT. KITCHEN - AFTERNOON

JAKE stands next to the sink reading a magazine.  Jake is
only seventeen but looks much older with his blue goatee and
two foot blue spiked Mohawk. He's sporting three earrings in
his left ear, five in his right and a shiny new one in his
left nostril.

His father, POPS, walks into the room dressed in a
conservative suit and looks much younger than his 54 years.
He puts his briefcase on the table.

                    POPS
                (excited)
          Your hair looks fabulous. Is that a
          new nose ring?

                    JAKE
          Gee thanks, Pops. I just got it
          yesterday. It was only ten thousand
          dollars.

                    MOM (O.S.)
          Dallas! You are not going to wear
          that suit again. It's embarrassing.
```

This format might look intimidating, but do not worry. Several inexpensive software packages that handle the formatting for you are available. Entire books have been written about formatting scripts, but they are obsolete now because of the abundance of script-formatting software.

After you know what your story is and how it is going to progress from beginning to end, it is time to actually write. The first step, as you learned, is determining the setting, and it is the first element you need to include on your script. Identify the location and whether it is outside (exterior) or inside (interior). You also need to identify the location every time there is a change.

Second, you need to write stage directions to tell actors what they are supposed to do when they are on camera. Be careful, however, that you do not dictate absolutely everything they do. Actors need to know what is happening in the scene and where they are supposed to be in relationship to everything else in the shot, but do not tell them when to breathe, for instance. Allow the actors (and the director) to interpret what is on the page. You will find that good actors can turn good writing into a great story.

Third, remember that the audience understands what is going on, so do not write dialogue that explains what the actors have just done. If an actor takes a drink and spits it out, do not write a line such as, "I have spit out my drink." Use action to show what characters do, and use dialogue to express what they feel. One thing about that, though, is that most emotions can be expressed through actions: Crying usually means somebody is sad. It is probably better to cut out as much of the dialogue as you can. Finally, follow the six style points from the section on writing news scripts. Read the script out loud, especially the dialogue. Listen to how people talk, and do your best do mimic the sound, feeling, and rhythm of real speech.

Now that you have learned how to develop a story, you can start writing the script in Final Draft. You will see quite a few similarities between Final Draft and Final Draft AV.

STEP-BY-STEP 2.7

In this Step-by-Step activity, you open Final Draft 7 and open a new document. Instructions for opening Final Draft 7 and a new document are provided for both Macintosh and Windows XP. Follow the instructions for your operating system.

Macintosh instructions

Follow these instructions if you are using a Macintosh:

1. To open Final Draft 7, click or double-click the **Macintosh HD** icon. (If the name of the hard drive has been changed, find and click it.)

2. Click or double-click the **Applications** icon in the frame on the left, or click or double-click the **Applications** folder in the second frame from the left.

3. Locate and click or double-click the **Final Draft 7** folder.

4. Click or double-click the **Final Draft 7** icon.

5. Final Draft creates a new script when you start the program. If not, use the following methods (if a New document dialog box appears with the stationery categories, click **Cancel**):
   a. At the left of the toolbar, click the New icon to open a new blank script.
   b. You can also click File, New from the menu.
   c. Press Command+N.

> **Hot Tip**
>
> As in Final Draft AV, you can place an alias (or shortcut, for those used to Windows terminology) on the Dock by dragging the icon onto the Dock. You also can click the icon, and then press Command+L to create an icon that you can drag to the desktop. To start the program, just double-click the icon on the desktop or Dock.

6. Before you proceed, take a quick look at the Final Draft layout. At the top is the menubar, which looks much the same as any menubar in Windows or Macintosh software. Just below that is the toolbar.

7. Next, look at the Element window, which tells you what element you are working on. You also can use it to navigate through the different elements. When you first start Final Draft, you should see "Scene Heading" in the Element window. See Figure 2-7.

8. Press **Tab** once and watch the Element window. The text changes to "Action." Press **Tab** again, and it changes to "Character." The Tab key allows you to navigate through the different screenplay elements.

9. Press **Tab** until you see "Scene Heading" in the Element window again.

10. Leave the program and document open for Step-by-Step 2.8.

STEP-BY-STEP 2.7 Continued

Windows XP instructions

Follow these instructions if you are using a PC:

1. Click **Start**, and point to **All Programs**.

2. Click the **Final Draft 7** icon. If an End User Agreement dialog box appears, click the **Accept** button to continue. If a Welcome to Final Draft 7 dialog box appears, click the Demo button to continue and if a Welcome to the Final Draft Demo dialog box appears, click **Continue**.

> **Hot Tip**
>
> You can create a shortcut for Final Draft by following Step 1, and then right-clicking the Final Draft 7 icon. On the shortcut menu, click Create Shortcut, and drag the new shortcut to the desktop.

3. Final Draft creates a new script when you start the program. If not, use one of the following methods to open a new script (if a New document dialog box appears with the stationery categories, click **Cancel**):

   a. At the left of the toolbar, click the **New** icon to open a new blank script.

   b. You also can click **File**, **New** from the menu; or

   c. Press **Ctrl+N** (Windows).

4. Before you proceed, take a quick look at the Final Draft layout, which is similar in both Windows and Macintosh.

5. At the top is the menubar, which looks much the same as any menubar in Windows or Macintosh software. Just below that is the toolbar.

6. Next, look at the Element window, which tells you what element you are working on. You also can use it to navigate through the different elements. When you first start Final Draft, you should see "Scene Heading" in the Element window. See Figure 2-7.

FIGURE 2-7
Element Window

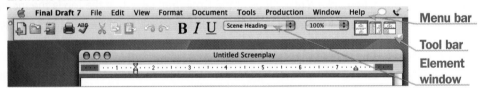

7. Press **Tab** once and watch the Element window. The text changes to "Action." Press **Tab** again, and it changes to "Character." The Tab key allows you to navigate through the different screenplay elements.

8. Press **Tab** until you see "Scene Heading" in the Element window again.

9. Leave the document and program open for Step-by-Step 2.8.

STEP-BY-STEP 2.8

In this step-by-step activity, you will create a script in Final Draft 7. The following steps work with both Windows XP and Macintosh. The figures in this Step-by-Step reflect both operating systems: The Macintosh figure is on the left and the Windows XP figure is on the right.

1. Make sure "Scene Heading" is showing in the Element window. If it has different text, press **Tab** until it does.

2. Type **E**. The "E" will appear with "XT." in gray as well as a small SmartType box. EXT. is highlighted in the SmartType box as shown in Figure 2-8.

FIGURE 2-8
SmartType Box (Macintosh on the left and Windows on the right)

3. When you write a script, you need to specify where the action is taking place, whether it is inside (interior, or INT.) or outside (exterior, or EXT.). Press **Tab** to accept EXT.

4. If you decide you want this scene inside, not outside, press **Delete** (Macintosh) or **Backspace** (Windows) five times. The cursor should move to the left of "EXT.," which appears in gray text. The SmartType box appears with the new choice highlighted. See Figure 2-9.

FIGURE 2-9
SmartType box with highlighted INT (MAC on the left and Windows on the right)

5. Press **Tab** to accept INT. So far the scene is inside, but you also need to indicate where the action takes place and what time of day it happens.

STEP-BY-STEP 2.8 Continued

6. Type **Kitchen** and press **Tab** to specify the location. At this point, a dash appears with "DAY" in gray and another SmartType box. See Figure 2-10.

FIGURE 2-10
SmartType Box specifying location

7. Press the **down arrow key** until AFTERNOON is highlighted, then press **Tab**. You have now identified when and where this first scene takes place.

8. Type **JAKE stands next to the sink reading a magazine. Jake is only seventeen but looks much older with his blue goatee and two foot blue spiked Mohawk. He's sporting three earrings in his left ear, five in his right, and a shiny new one in his left nostril.**

9. Press **Tab**. The Element window has not changed because anything can follow an action: more action, dialogue, a scene heading, and so forth. You are going to stay with action and type **His father, POPS, walks into the room dressed in a conservative suit and looks much younger than his 54 years. He puts his briefcase on the table.** See Figure 2-11.

Important

Notice that the heading in the Element window has changed from "Scene Heading" to "Action." Remember, Final Draft moves through the elements according to what usually comes next, so after a scene heading you typically go to an action (also called a "stage direction").

Important

When you introduce a major character, put the name in uppercase letters (but only the first time). You also need to state what the character looks like, how old the character is, and other details of the character's physical appearance.

STEP-BY-STEP 2.8 Continued

FIGURE 2-11
Element Window showing Action

10. Press **Tab** twice. Notice that the cursor has moved to the center of the line and the Element window now reads "Character." See Figure 2-12.

FIGURE 2-12
Element Window reads character

11. Type **pops**. Final Draft puts the character's name in uppercase letters automatically, so do not try to change it.

12. Press **Tab**. The cursor moves a little to the left below "POPS" and is surrounded by parentheses. Inside the parentheses, you can add directions for how the line should be delivered. Type **excited** and press **Tab**. The Element window now reads "Dialogue," and the cursor has moved down and a little farther to the left. See Figure 2-13.

FIGURE 2-13
Element Window reads Dialogue

13. Type **Your hair looks fabulous. Is that a new nose ring?** Press Tab.

> **Important**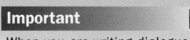
>
> When you are writing dialogue, you need to give actors enough information about the characters so that they can play their parts, but not too much detailed that they cannot make the part their own. Video and film are collaborative arts, so let other people, like the actors, give their input.

STEP-BY-STEP 2.8 Continued

14. Press **Enter** (Windows) or **Return** (Macintosh) to see a pop-up list of screenplay elements. See Figure 2-14.

FIGURE 2-14
Pop-up list of screenplay elements

15. Press the **Up arrow key** until Character is selected. Press **Enter** or **Return**. Type **Jake** and press **Enter** or **Return** to bypass the parenthetical direction and go to dialogue.

16. Type **Gee thanks, Pops. I just got it yesterday. It was only ten thousand dollars.** Press **Return** or **Enter**. The Element window changes to "Action," so press **Tab**. The cursor moves back to the center and "POPS" appears in gray as shown in Figure 2-15. Final Draft assumes you want to write a conversation between the two characters.

FIGURE 2-15
Cursor moves to the center and POPS appears gray

17. Type **mom** to add another character to the scene. Press the **spacebar** and type ((a open parenthesis). A new SmartType box appears. The first item in the list is (V.O.) for voiceover. See Figure 2-16.

STEP-BY-STEP 2.8 Continued

FIGURE 2-16
Voice-over element

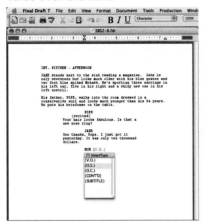

18. Press the **down arrow key** until (O.S.) is selected, and press **Tab**. Mom is in the scene but off screen, so the audience does not see her. The characters can hear her and can react to whatever she says.

19. Press **Return** or **Enter**. You are now ready to type Mom's dialogue. (If you had pressed Tab instead, you could have typed a parenthetical direction.)

20. Type **Dallas! You are not going to wear that suit again. It's embarrassing.** Press **Return** or **Enter**. See Figure 2-17 for an example of how your script should look.

FIGURE 2-17
Example of fictional narrative script

21. Save the file as **SBS2-8** and close the program.

Final Draft is a simple program, but it does take some getting used to. Other software packages for writing screenplays are available, but the script looks similar no matter what software you use. People working on a film or video expect the script to be formatted a certain way. If the script does not look professional, the project will not be taken seriously.

One more thing—Make sure you write a script that's practical to shoot. It will not do you any good to write a script with a 50-car pile up if you cannot get 50 cars to wreck. Limit yourself to what is available to you.

Anyone can learn to write well with practice and attention, so putting in the time is well worth it. Scripting is just the first part of preproduction, however. In the next lesson, you learn how to breakdown the script so you can finish up preproduction.

SUMMARY

In this lesson, you learned:

- The elements of a story are conflict, caring about the story, and getting to the point.

- The story structure of a fictional narrative is a clear beginning with an protagonist and an antagonist, a middle with an external and internal conflict, and the end where the climax is reached.

- Content (what you write) and style (how you write) are important in writing a script.

- You can use software, like Final Draft AV 2.5 and Final Draft 7, to write a news script, a fictional narrative script, and a documentary script.

- The six points of style in writing a news broadcast:

    a) Write in a conversational style

    b) Use active sentence

    c) Start strong

    d) Write simply

    e) Do not be wordy

    f) Write to be spoken

- There are two approaches to the same story: news vs. opinion.

- There is a difference between information and propaganda in creating a documentary.

VOCABULARY *Review*

Define the following terms:		
Active sentence	External conflict	Protagonist
Antagonist	Internal conflict	Script
Conflict	Narrative	Story
Content	Passive sentence	Style
Documentary		

REVIEW *Questions*

WRITTEN QUESTIONS

Write a brief answer to the following questions.

1. What are the three things a storyteller knows?

2. What is conflict?

3. Describe the six points of style for clear writing.

4. Explain the process of writing a script for a documentary.

5. Describe the development of a narrative story.

PROJECTS

PROJECT 2-1

Watch a movie and pay attention to the points covered in this lesson. Identify the protagonist and his or her quest. Identify the conflict and what is keeping the protagonist from achieving his or her goal. Identify the obstacles the protagonist faces and how he or she reacts to them. Identify the climax and how the story reaches its conclusion. Develop a short narrative using the process covered in Step-by-Step 2.6.

PROJECT 2-2

Write a script. You can write a news script, a fictional narrative, or a documentary for a video that is 5 to 10 minutes long. The rule of thumb is that one page of script should equal one minute of screen time, so if you are writing the script for a 5- to 10-minute video you need to write 5 to 10 pages. Remember what you have learned about good storytelling. It might seem hard to put all the storytelling elements in a 5- to 10-minute script, but it can be done. Consider 10-minute cartoons.

PROJECT 2-3

Pick a story from a newspaper or magazine and rewrite it for broadcast.

PROJECT 2-4

Pick a subject. It can be a family member, your best friend, or a topic you are interested in, such as World War II or fly fishing. Write an outline, following the steps you used earlier. Can you identify a conflict? Can you see why somebody else would want to watch your story? Is your story focused?

 ## WEB PROJECT

Research the Internet for good techniques in writing a script for either a news broadcast, documentary, or fictional narrative. Write down your research and present it to the class. Compare and contrast the research others have done.

 ## TEAMWORK PROJECT

Work together in groups of three and create a script for any of the three types of production you learned about in this lesson. After you finish writing your script, have another group critique the script according to the elements of a story and six points of style. Keep in mind the difference between news vs. opinion and information vs. propaganda. If writing a documentary, follow the steps outlined in the documentary section. If writing a fictional narrative, follow the classic story structure.

CRITICAL *Thinking*

ACTIVITY 2-1

Watch the news. Listen to the anchors and pay attention to the reporters. How do they sound? What do they say? How do they say it? What do you think they could do to improve how they present their stories? One of the best ways to learn how to do this is watch and listen to somebody else. Pay attention and learn!

ACTIVITY 2-2

Watch or rent a documentary. PBS, History Channel, and Discovery Channel are full of documentaries on a variety of topics. Pay attention to the focus of the story. Does the documentary cover the entire subject, or does it focus on one particular aspect? Are you engaged in the story? Why or why not?

SCRIPT BREAKDOWN

OBJECTIVES

Upon completion of this lesson, you should be able to:

- Use Final Draft to create a shooting script
- Break down a narrative script
- Use Word to create a scene breakdown sheet
- Break down a documentary script
- Mark a multi-camera script

Estimated Time: 1.5 hours

VOCABULARY

Cast

Extras

Production notes

Props

Scene breakdown sheet

Scene numbers

Script breakdown

Shooting script

Sound effects/music

Special effects

Special equipment

Stunts

Vehicles/animals

Introduction

You have your script and you are ready to start shooting, right? Not quite, you are not quite finished with preproduction. While what we cover in this lesson is not as exciting or creative as script writing or shooting and editing, it is just as important. Production runs smoother and is more enjoyable if you take the time to prepare.

Before you can put your script into production, you need to finish the preproduction stage. This phase of preproduction is the office work. Office work includes the script breakdown, budgeting, scheduling, casting, location scouting, and storyboarding. The first element in that list, the script breakdown, is the basis for all of the other office work.

The script is the guide for the story, but the script breakdown is the guide for production. The breakdown helps you identify what casting needs to be done, what locations to plan on, and how much money is needed and where it is spent. The script breakdown may even help you decide to rewrite parts of the script. The budgeting, scheduling, casting, location scouting, and storyboarding can happen in almost any order, but the script breakdown has to happen first. This lesson covers script breakdown.

Using Final Draft to Create a Shooting Script

In order to make the script breakdown easier for the production, there are two documents that you create: first a shooting script and then a scene breakdown sheet. Many of the scripts you can check out from the library are shooting scripts. Shooting scripts include scene numbers and revisions. You do not want to create a shooting script if you are trying to sell the script to somebody else, like a Hollywood director. Trying to sell a shooting script makes you look like an amateur. Do not waste your time. But, if you write the script and plan to shoot it, you will need to make a shooting script.

Transitions or camera angles can be added to a shooting script along with revisions. The most important part of a shooting script, however, is the scene numbers. Numbering the scenes helps keep track of each scene and communicates with others about the specific scene. If you make a lot of revisions to a script the page numbers will likely change, but the scene numbers will not. The scene numbers are also an important part of the scene breakdown sheet. You use the scene numbers from the shooting script in the scene breakdown sheet. Creating a **scene breakdown sheet** identifies the individual elements needed to shoot each scene. It includes what needs to be shot, where things are going to be shot, and everything that needs to be included in each scene.

> **Note**
>
> You can add transitions or camera angles to the script, but they are not necessary. I never add transitions when I write or shoot a script, and I only add camera angles when I need to be specific about a shot, like a close-up on a piece of machinery.

STEP-BY-STEP 3.1

In this Step-by-Step activity, you turn your script into a shooting script by adding scene numbers. The steps are the same for both MAC and PC unless otherwise indicated. Screen shots reflect the Final Draft 7.0 for Macintosh.

1. Launch **Final Draft 7.0** by clicking or double-clicking the Final Draft 7 icon.

2. Open the file named **SBS 2-8.fdr (MAC) or SBS2-8W.fdr (PC)** by clicking **File**, and then clicking **Open** from the menubar and then double-click the file.

3. Click **Production**, and then click **Scene Numbers** from the file menu.

4. Click in the checkbox next to **Number/Renumber** (if you are working with a MAC) or Add Numbers (if you are working with a PC). Make sure it says Scene Heading in the element drop down menu. If anything else is in the element window, click the arrows to the right and select Scene Heading as seen in Figure 3-1.

> **Note**
>
> You also can open your script by clicking **File**, then clicking **Open** or **Open Recent** (depending on the version of Word you are using), and selecting the script name.

STEP-BY-STEP 3.1 Continued

FIGURE 3-1
Scene Numbers dialog box

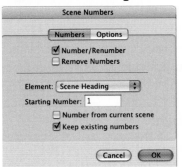

5. Click **OK**. Numbers appear next to each scene heading as seen in Figure 3-2.

FIGURE 3-2
Script with scene number added

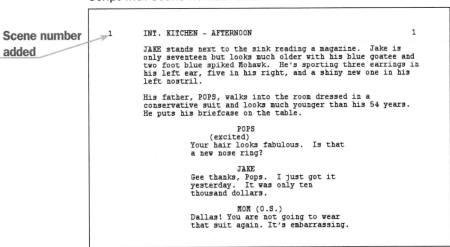

6. Save the script as **SBS3-1.fdr**. Your script should look like the one in Figure 3-2. Close the document and exit the program.

Now that you have a shooting script you can start the breakdown. Many of the steps we covered for a narrative script breakdown are used for both documentary and news scripts. Each type of script, however, has a few unique elements that we will cover.

Break Down a Narrative Script

We will start with the narrative script since it is probably the most involved and time consuming.

Most people are surprised when they find out that movies usually are not shot in order. The first scene is not always the first scene shot, and the last scene is not always the last scene shot. If the first and third scenes take place in the same detective's office, and the second scene takes place at a crime scene, it does not make sense to shoot the first scene on Day 1, the second scene on Day 2, and the third scene on Day 3. If you shoot the scenes in order, it is very expensive to move people and equipment from once place to another and you spend the entire budget on gas and travel time. For example, if you have an actor that is only seen in the detective's office scenes and you shoot those scenes on Day 1 and Day 3, you may have to pay the actor for Day 2 because it is often hard for the actor to find work on the off day.

Doing a script breakdown basically helps you identify the individual elements needed to shoot each scene as discussed earlier. Breaking down a script actually is breaking down each scene, so you will need a scene breakdown sheet.

Use Word to Create a Scene Breakdown Sheet

The first thing we are going to do is to create a scene breakdown sheet using Microsoft Word. I have seen a lot of different breakdown sheets, there is not an "official" one. You can lay out your breakdown sheet any way you like. Make sure to include the Scene Breakdown Elements covered in this lesson so it is easy for the production crew to understand.

S TEP-BY-STEP 3.2

In this Step-by-Step activity, you create a header for the breakdown sheet.

1. Start Microsoft Word.

2. You should see a new, blank document when you start Word. If you do not have a new, blank document, click **File**, then click **New** from the menubar.

3. Click **Table**, point to **Insert**, and then click **Table** to open the Insert Table dialog box.

4. Click in the Number of columns box and highlight the existing number, if necessary, then and type **3**. Click in the Number of rows box, highlight the existing number, and type **4**.

5. Click **OK**. You should now have 4 rows of cells in 3 columns. The cursor should be in the first cell on the left.

6. Type **Title:**. This is where you put the title of the video you are working on.

7. Click and drag the mouse over the first two cells to select them. (See Figure 3-3)

FIGURE 3-3
Selecting table cells

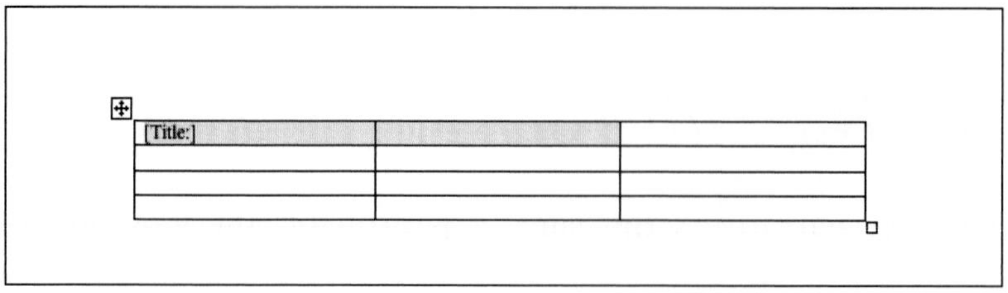

STEP-BY-STEP 3.2 Continued

8. Click **Table**, then click **Merge Cells** from the menubar. This merges the two cells into a single cell. Press **Tab** to move to the next cell.

9. Type **Breakdown Sheet No.** You want to number the breakdown sheets to keep them in order. Press **Tab** to move to the next cell.

10. Type **Scene Number:**. Press **Tab** to move to the next cell.

11. Type **Script Pages:**. This gives you two ways to identify each scene. Press **Tab** to move to the next cell.

12. Type **Date:**. You want to make sure you know when the scene breakdown was completed so you do not miss revisions. Press **Tab** to move to the next cell.

13. Type **Location:**. Click and drag over the cells in this row, including the Location cell.

14. Click **Table**, then click **Merge Cells** from the menubar. This merges the row into a single cell. Press **Tab** to move to the next cell.

15. Type **Page Count:**. Each scene will be divided into eighths for scheduling purposes. Press **Tab** to move to the next cell.

16. Press **Caps Lock**. Type **EXT/INT**. This refers to the interior or exterior that you wrote in a script location. Press **Tab** to move to the next cell.

17. Type **DAY/NIGHT**.

18. Press the down arrow key to move the cursor out of the table. Press **Return** or **Enter**. This creates space between the header you just created and the next table you are going to create. Your breakdown sheet should look like Figure 3-4.

19. Save the document as **SBS3-2.doc** and leave the program and document open for the next Step-by-Step activity.

FIGURE 3-4
Header of a breakdown sheet

Title:		Breakdown Sheet No.
Scene Number:	Script Pages:	Date:
Location:		
Page Count:	EXT/INT	DAY/NIGHT

So far we have created the header for the breakdown sheet. The header helps you keep the scenes organized. The next section introduces the elements that are included in a breakdown sheet.

Scene Breakdown Elements

The most common system for a script breakdown is to color code each element. For example, the name of every character that speaks in a scene is underlined with a red colored pencil. The purpose for color coding a script breakdown is to identify the elements that are transferred to the scene breakdown sheet. It is a quick way for the director and production crew to identify what is needed for a specific scene.

The information you need to include in the breakdown sheet is as follows:

- **Cast:** This is any character that speaks. Underline their name in red the first time their name appears in a scene, usually in the scene descriptions. You may have people say something that is not important to the story or are never heard from or seen again, they are considered extras.

- **Special Effects:** This is anything like a spaceship moving through a galaxy or a light saber. Underline these in blue.

- **Wardrobe:** This is any specific clothing a character wears. Circle the wardrobe with a number 2 pencil or with a pen.

- **Special Equipment:** This is any specialized equipment that is used to shoot the scene. Things like a dolly or crane go in this cell. Draw a box around any equipment needed with a pen. You usually would see something like dolly shot or crane shot in the script.

- **Stunts:** This is any kind of dangerous physical action, like a fall or jumping from one building to another, which requires special safety precautions. Underline these with orange.

- **Extras:** These are people that are not main characters and usually do not deliver a line, like in a street scene where people walk by or around the main character. Underline these in green.

- **Props:** Props are anything that is handled by a character, like a book, computer, or even a shoe. Underline these with purple.

- **Make-up/Hair:** Use an asterisk to identify these elements. This refers to any kind of physical change that needs to be made to an actor. It may include a two foot blue Mohawk or a scar.

- **Production Notes:** This is anything else, like any questions, about a scene. Any type of information that is unclear in the script should be underlined in black and the question written in the breakdown sheet.

- **Vehicles/Animals:** This refers to animals or any vehicles that are in a scene, like a car driving by one of the characters, or a dog barking. Underline these with pink.

- **Sound Effects/Music:** This is any specific songs or music that is playing in the scene (not the film score), or any extra sounds, like a gunshot or a door slamming. These usually are created in post-production, but need to be identified in the breakdown to ensure they are included. Identify these in brown.

S TEP-BY-STEP 3.3

In this Step-by-Step, you add some of scene breakdown elements to your breakdown sheet.

1. With Microsoft Word open and **SBS3-2.doc** opened, click **Table**, then point to **Insert**, and then click **Table** from the menubar.

2. Type **3** in the Number of columns box, type **5** in the Number of rows box, and then click **OK**. This will create a new table.

3. Type Cast in the first cell. Press **Enter** or **Return**.

4. Type **(Red)**.

5. Fill out the rest of the table so it matches the table in Figure 3-5. Make sure to include the color code in parentheses.

6. Move the cursor over the bottom line of the table until the cursor becomes two lines with an up arrow and a down arrow.

7. Click and drag the last row of cells to the bottom of the page. Do not go too far, this will make the bottom row go onto the next page.

8. Select the entire table, then click **Table**, point to **Autofit**, and then click **Distribute Rows Evenly**.

9. On the table, merge the Cast cell with the cell below it, merge the Extras cell with the cell below it, and merge the Production Notes: cell with the two to the right of it. Your Breakdown sheet should fit on a single page and look like the one in Figure 3-5.

STEP-BY-STEP 3.3 Continued

FIGURE 3-5
Adding elements to the breakdown sheet

Title:		Breakdown Sheet No.
Scene Number:	Script Pages:	Date:
Location:		
Page Count:	EXT/INT	DAY/NIGHT

Cast (Red)	Stunts (Orange)	Extras (Green)
	Sound Effects/Music (Brown)	
Special Effects (Blue)	Props (Purple)	Vehicle/Animals (Pink)
Wardrobe (Circle)	Make-up/Hair (Asterisk)	Special Equipment (Box)

Production Notes: (Underline)

10. Save the file as **SBS3-3.doc** and leave the document open for the next Step-by-Step activity.

Now that you have your breakdown sheet it's time to break down the script.

STEP-BY-STEP 3.4

In this Step-by-Step activity, you are going to fill out your scene breakdown sheet using the narrative script you wrote in Lesson 2. You also mark the scene elements in color.

1. Start **Final Draft 7**, then open and print out **SBS2-8.fdr** or **SBS2-8W.fdr**.

2. With **SBS3-3.doc** open, type **Jake's Dad** in the Title: cell. Type **1** in the Breakdown Sheet No: cell.

3. Type **1** in the Scene Number: cell. Type **1** in the Script Pages: cell.

4. Type today's date in the Date: cell.

5. Type **Kitchen** in the Location: cell in the header.

STEP-BY-STEP 3.4 Continued

6. Using the printout of SBS2-8.fdr or SBS2-8W.fdr, break down each scene into eighths. An eighth is equal to about an inch of page. Use a ruler to mark one inch section in the right margin on the script, and then write **1/8** next to the marks. (See Figure 3-6)

FIGURE 3-6
Dividing the script in eighths

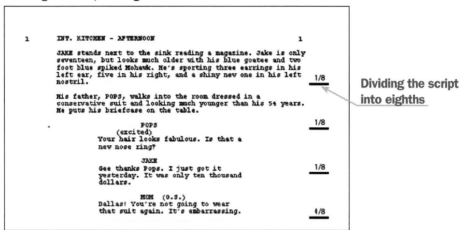

7. Count the approximate number of eighths in the scene. There are four eighths in scene 1.

8. In SBS3-3.doc document, type **4/8** in the Page Count cell. This will help you with scheduling later on.

9. In EXT/INT cell, underline **INT** and in the DAY/NIGHT cell, underline **DAY**. Your breakdown sheet should look like Figure 3-7.

FIGURE 3-7
Completed header of the breakdown sheet

Title: Jakes's Dad		Breakdown Sheet No. 1
Scene Number: 1	Script Pages: 1	Date: [today's date]
Location: Kitchen		
Page Count: 4/8	EXT/INT	DAY/NIGHT

10. Using the printout of SBS2-8.fdr, read through the first paragraph of scene description. Underline **Pops and Jake** with red.

11. On the SBS3-3.doc document, Type **Jake**, press **Enter**, then type **Pops** in the Cast cell.

12. Continue reading the script. Underline magazine and briefcase in purple, transfer this information to SBS3-3.doc and type **Magazine** and **Briefcase** in the Props cell.

> **Note** ☑️
>
> Transfer as you go, not when you are finished marking the scene. Missing something, like the briefcase, can cause a lot of problems when you are shooting. Do not make this mistake.

13. On the script printout, circle **three earrings in his left ear**, **five in his right**, **a shiny new one in his left nostril**, and **conservative suit**. Transfer this information to SBS3-3 and type **Nose ring, Jake; 8 earrings, Jake; Conservative suit, Pops** in the Wardrobe cell.

STEP-BY-STEP 3.4 Continued

14. On the script printout, place an asterisk next to **blue goatee** and **two foot blue spiked Mohawk**. Transfer this information to SBS3-3 and type **Blue Mohawk, Jake**; and **Blue Goatee**, **Jake** in the Make-up/Hair cell.

15. When you have finished marking the script (see Figure 3-8), compare it to the finished breakdown sheet (See Figure 3-9). If you missed something, go back and figure out what you missed.

> **Note**
>
> There is software that does all of the marking for you, but some of it is really expensive. Using software, however, does not make you think about what is actually going on in each scene and what it takes to shoot everything. Learn to break down a script "by hand" first, then when you have the software, you will know exactly what is going on and will be able to identify any mistakes or problems.

FIGURE 3-8
Marked script

```
1      INT. KITCHEN - AFTERNOON                              1

       JAKE stands next to the sink reading a magazine. Jake is only
       seventeen, but looks much older with his blue goatee and two
       foot blue spiked Mohawk. He's sporting three earrings in his
       left ear, five in his right, and a shiny new one in his left
       nostril.                                                       1/8

       His father, POPS, walks into the room dressed in a
       conservative suit and looking much younger than his 54 years.
       He puts his briefcase on the table.
                                                                      1/8
                           POPS
                        (excited)
                Your hair looks fabulous. Is that a
                new nose ring?

                           JAKE
                Gee thanks Pops. I just got it                        1/8
                yesterday. It was only ten thousand
                dollars.

                           MOM  (O.S.)
                Dallas! You're not going to wear
                that suit again. It's embarrassing.                   1/8
```

STEP-BY-STEP 3.4 Continued

FIGURE 3-9
Finished breakdown sheet

Title: Jake's Dad		Breakdown Sheet No. 1
Scene Number: 1	Script Pages: 1	Date: [today's date]
Location: Kitchen		
Page Count: 4/8	EXT/<u>INT</u>	<u>DAY</u>/NIGHT

Cast (Red) Jake Pops	Stunts (Orange)	Extras (Green)
	Sound Effects/Music (Brown)	
Special Effects (Blue)	Props (Purple) Magazine Briefcase	Vehicle/Animals (Pink)
Wardrobe (Circle) Nose ring, Jake 8 earrings, Jake Conservative suit, Pops	Make-up/Hair (Asterisk) Blue Mohawk, Jake Blue Goatee, Jake	Special Equipment (Box)
Production Notes: (Underline)		

16. Save the file as **SBS3-4.doc** and close the document. Exit Microsoft Word.

Congratulations. You have just broken down your first narrative script.

Break Down a Documentary Script

Breaking down a documentary script is much the same as breaking down a narrative script: read the script and identify what footage is needed to complete the script. You can even use the same scene breakdown sheet you created earlier. The main objective here is to identify what you need to complete the video. Some of the footage may be still photographs, existing videos, or movies. That type of footage is great to use because it is already in existence. The only problem may be getting it from whatever format it is in, such as a still photograph that needs to be scanned, to a video format you can work with.

Instead of casting, however, you may need to identify people you will interview. You also may need to identify some one to do the narration, called a voice over, so you can cast and schedule voice talent and an audio booth. Location scouting also may be the same as identifying people to interview because you need to know where you will interview people, or, if the documentary is about a place, you may want to shoot video of the actual place. Probably the best way to understand how all this works is to take a look at a documentary script and do a simple breakdown.

S TEP-BY-STEP 3.5

In this Step-by-Step activity, you create a scene breakdown sheet for a documentary script that you created in Lesson 2. You actually are going to create three scene breakdown sheets for this activity.

1. Open Final Draft AV 2.5 and then open **SBS2-5.xav** or **SBS2-5W.xav**. Open **breakdownsheet.doc**.

2. To fill out the header for the first scene, type the following into the header:
   a. In the Title cell, type **Grandma's house**.
   b. In the Breakdown Sheet No. cell, type **1**.
   c. In the Scene number cell, type **1**.
   d. In the Script Pages cell, type **1 of 1**.
   e. In the Date cell, type [today's date].
   f. In the Location cell type **Grandma's house from the front yard**.
   g. In the Page Count cell, type **3/8**.
   h. In the EXT/INT cell, bold **EXT**. In the DAY/NIGHT cell, bold **DAY**.

3. Read the first line under VIDEO on the script. The first line says Grandma's house from the front yard. On the script, underline Video of Grandma's house from the front yard. In the Production Notes cell of the breakdown sheet, type **Video of Grandma's house from the front yard**.

4. Read the audio information on the script. It says the narrator has a short sentence. On the script, underline NARRATOR in red by clicking **Format**, then **Set Font**. The word Narrator will also be in red. In the Cast cell of the breakdown sheet, type **Narrator**. You have completed the first breakdown sheet.

5. Save the script and the breakdown sheet as **SBS3-5a.doc** and leave them open. Continue to the next step.

6. Create a new header for the second breakdown sheet. Type the following into the header:
   a. In Scene Number cell, type **2**.
   b. In the Breakdown Sheet No. cell, type **2**.
   c. In the Location cell, type **Inside Grandma's house: the bathroom, the kitchen, and the livingroom**.
   d. In the EXT/INT, bold **INT**.
   e. Leave the rest of the cells the same.

7. Read the next section under video on the script. It says Montage of Grandma's house, the bathroom, the kitchen, her television. On the script, underline Montage of Grandma's house: the bathroom, the kitchen, her television. In the Production Notes cell of the breakdown sheet, type **Video montage of the inside of grandma's house: her bathroom, her kitchen, her livingroom**.

STEP-BY-STEP 3.5 Continued

8. On the script, read the second section under Audio. It has more lines for the narrator; you only need to mark the first occurrence of a cast member. On the script, do not underline Narrator in red. However, in the Cast cell of the breakdown sheet, verify that Narrator is in the Cast cell of the breakdown sheet. You have completed the second breakdown sheet.

9. Save the script and breakdown sheet as **SBS3-5b.doc** and leave them open. Continue to the next step.

10. To create a new header for the third breakdown sheet, type the following into the header:

    A. In the Breakdown Sheet No. cell, type **3**.

    B. In the Scene Number cell, type **3**.

    C. In the Location cell, type **Inside grandma's house**.

    D. Leave the rest of the header the same.

11. On the script, read the third section under video. Underline Still images of house in Yukon. In the Production notes cell of the breakdown sheet, type **Still image of house in Yukon**.

12. On the script, read the third section under audio. Underline Grandma and Me in red. In the Cast cell of the breakdown sheet, type **GRANDMA** and **ME**. In the Production Notes cell, type **Do a video shoot with Grandma and me**. You have completed the third breakdown sheet.

13. Save the script and breakdown sheet as **SBS3-5c.doc** and close the script, breakdown sheet, and the applications of Final Draft AV and Microsoft Word.

    You should end up with three breakdown sheets, one for each scene, that identifies what elements you need to put the script together. Breaking down a documentary script may seem like a waste of time but it helps you in the long run. More than once I have run into a problem putting a script together because I did not take the time to break the script down and identify what I needed.

Mark a Multi-Camera Script

Breaking down a multi-camera shoot can be very similar to the breakdowns for the other shoots we have just covered. You need to scout for locations, identify a cast, figure out props, etc. In this section, however, I make the assumption that location has been determined, the casting has happened, and what you are shooting is a repeatable event, such as a daily newscast. This section shows the director how to mark a script. The director of a news broadcast or any scripted shoot marks the script so they know what each camera shoots and when it happens. In live news, for example, you have a specific amount of time for the shoot. If you go over or under

> **Note**
>
> You already may have your location for a sporting event, such as a football stadium, but that does not mean you do not need to scout the location before you get there. There is nothing worse than showing up and setting up your camera only to find that the twenty-five foot extension cord you brought is too short. Make sure to go to the field or court and know the location of the power outlets. Make sure you know how to get to camera positions so you can get in and set up quickly. Find out when the doors open and when the crowd gets there. This information helps you determine when you need to be there and what kind of equipment you need.

that time, you get into trouble. The idea is to get everything prepared before hand so you know if you need to adjust accordingly. Of course there are always problems, but if you plan ahead you should be able to avoid these problems.

The challenge of a live broadcast is that things go wrong. You may plan on using all three cameras, but suddenly find out only two of the three cameras are working. You have to be flexible. Marking your script allows you to make decisions before you are on air when there is little or no pressure. This allows you to make fewer decisions when you are on air when there is a lot of pressure.

First, identify which camera is used to cover which story. This decision depends on which anchor is taking which story. Each camera should be numbered so it can easily be identified. Camera 1 should be on a close-up of the same anchor the entire broadcast, but you probably want to go to a full shot of the set between the news and the sports to make the transition smoother. Write the camera number along the right-hand side of the script.

Second, identify which tapes are in which machines and when those tapes need to be played. When the script calls for the highlights from last night's soccer game, the director needs to know which tape the highlights are on, and then which tape machine that tape is in. Marking the script helps the director know when the sports anchor expects the highlights to come up, how long the highlights last, and when to go back to the live camera. This is all very tricky, but careful thought and planning beforehand helps things run smoothly.

Third, identify graphics. The director needs to work with the graphics operator to make sure they are on the same page as to which graphic is shown. The graphics should be created beforehand as much as possible, although the graphics person needs to be able to create new graphics on demand.

Fourth, mark timing. For example, when the sports anchor introduces a highlights package the director needs to know when the highlights are over. The director also needs to find out how long the highlights run so she knows when they are finished. This allows the director to tell the sports anchor when they are live on the air again. Otherwise, the anchor may be scratching their nose, or something worse, when they are on camera again.

Marking scripts for other types of live broadcasts can be more involved, like doing a live broadcast of a drama or sitcom. The same principles still apply.

There are a couple more things on marking a script for news broadcast. First, make sure to keep marking to a minimum and keep the marking clear. The script should not be covered with notes that are hard to read. The director needs to be able to watch what is happening and pay attention to how things are going. The director will not be able to do that if they are trying to read a ton of notes.

Second, when you mark a script, make the marks in the same location so you do not have to look all over the place for what you have written down. For example, mark camera decisions along the right-hand side of the script.

> **Note**
>
> You can determine which camera shows what and when for a newscast, but not for a sporting event. You can mark up the script for a pre- or post-game show, but not for what happens during the game. If you are lucky enough to be able to do a sporting event with multiple cameras you need to watch and pay attention so you have the right camera getting the action at the right time. It takes practice, but you can learn how to shoot live, un-scripted events and do very well.

STEP-BY-STEP 3.6

In this Step-by-Step activity, you mark up a newscast. There are two cameras for this newscast, one computer graphic, and one tape that lasts 30 seconds to be played on video tape recorder X.

1. Open **newsscript.pdf** from your data files folder and print it out.

2. In the right margin of the script, write in red **cam 3** next to the first Jason line. Write in red **cam 2** next to the third Jason line. (See Figure 3-10)

FIGURE 3-10
Marked up newscast script

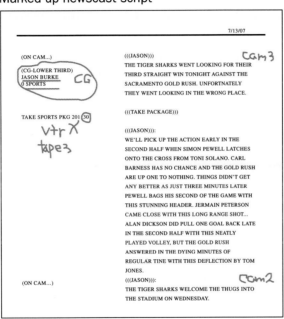

3. To mark the computer graphic of Jason Burke, 9 Sports, circle in red **(CG-LOWER THIRD)**, **JASON BURKE**, **9 SPORTS** in the left column, then write in red **CG** inside the circle, and underline in red **Jason Burke, 9 Sports** as seen in Figure 3-10.

> **Note**
>
> The CG-LOWER THIRD means computer graphic in the lower third of the monitor.

4. To mark for Tape 3 in Video Tape Recorder X, write in red **VTR X Tape 3** beneath **TAKE SPORTS PKG 201:30** as seen in Figure 3-10.

5. To mark the timing of Tape 3, locate **TAKE SPORTS PKG 201:30**. Circle in red the **:30**. This means that the tape will take 30 seconds to play as seen in Figure 3-10.

> **Note**
>
> Roll Tape 3 is easier for the director to say than roll Take Sports Pkg 201:30 from VTR X.

6. Submit the marked up script to your instructor.

SUMMARY

In this lesson, you learned:

- The script is the guide for the story; the script breakdown is the guide for production.

- A shooting script consists of scene numbers and revisions. It is used to create the scene breakdown sheet.

- A script breakdown identifies what needs to be shot, where things are going to be shot, and all the individual elements needed to shoot a scene.

- Cast, Special Effects, Wardrobe, Special Equipment, Stunts, Extras, Props, Makeup/Hair, Production Notes, Vehicles/Animals, and Sound Effects/Music are common elements found in a scene breakdown sheet.

- Breaking down a documentary script is much the same as breaking down a narrative script.

- Identifying the camera, tapes, tape machines, graphics, and timing are part of marking a multi-camera script.

VOCABULARY *Review*

Define the following terms:		
Cast extras	Scene numbers	Special effects
Production notes	Script breakdown	Special equipment
Props	Shooting script	Stunts
Scene breakdown sheet	Sound effects/music	Vehicles/animals

REVIEW *Questions*

TRUE/FALSE

Circle T if the statement is true or F if the statement is false.

T   F   1. Each camera operator in a multi-camera shoot determines what and when they will shoot.

T   F   2. A documentary script does not need to be broken down.

T   F   3. It is a good idea to add scene numbers to a script you are hoping to sell.

T   F   4. Scheduling is more important that script breakdown.

T   F   5. The director of a live shoot is responsible to keep track of everything that is going on in the shoot.

T   F   6. There is only one color code for a script breakdown.

T   F   7. You should color code the entire scene before you transfer the information to the breakdown sheet.

T   F   8. You need a new breakdown sheet for each scene.

MULTIPLE CHOICE

Select the best response for the following statements.

1. What are Props referring to?
   a. anything seen on the screen
   b. the location (property) where you are shooting
   c. any item handled by the cast on screen
   d. any item owned by a production company

2. What are Stunts referring to?
   a. any action on screen
   b. dangerous physical actions
   c. car chases only
   d. explosions

3. What is Sound effects/music referring to?
   a. any sound heard at any point
   b. the musical score
   c. all background sounds on set
   d. any extra sounds, usually added in post-production

4. What does Cast refer to?
   a. main characters only
   b. supporting characters only
   c. the antagonist and protagonist
   d. any characters seen on screen

5. What are Special effects?
   a. the same as stunts
   b. only elements created by computer
   c. created in post-production
   d. actions that require equipment other than a camera alone

WRITTEN QUESTIONS

Write a brief answer to the following questions.

1. Explain the purpose of a script breakdown.

2. Explain the purpose of color coding a script breakdown.

3. Describe the four elements needed to mark a news script.

PROJECTS

PROJECT 3-1

1. Open **Project 2-2**. You created this project in Lesson 2.

2. Create a shooting script for the script by adding scene numbers.

3. Create a scene breakdown sheet and print out enough copies for each scene in your script.

4. Break down each scene in your narrative or documentary script. Mark your script and transfer the information onto the breakdown sheet for each scene.

5. If this is a multi-camera script, then mark the camera numbers, names of the tape machines and tape numbers, computer graphics, and timing of the tapes.

6. Save your file as **Project 3-1** in your solution files folder.

 WEB PROJECT

Search the Internet for different script breakdown sheets and discuss the differences, similarities, advantages, and disadvantages of each one. Report your findings to the class in an oral or written presentation.

 TEAMWORK PROJECT

Get with a couple of classmates, or have your teacher divide the class into groups. Have the group create a script. Have each student do a breakdown or mark the script. How is each breakdown or script the same? How is each one different? Are the differences significant? How would the differences change the shoot? What could each member of the group do to improve their work?

CRITICAL *Thinking*

ACTIVITY 3-1

Take a look at the video breakdown sheet you have created. What kind of sheet did you create? What would you keep, and what would you do differently? What would a breakdown sheet for a documentary look like? Design a new sheet that fits your needs. What does it look like? What are the differences and the similarities? Why did you create it the way you did and how does it improve what you do?

STRIP BOARDING, SCHEDULING, AND BUDGETING

Introduction

So far, you should understand the basic concepts of both analog and digital video. In Project 2-2, you created a script for a 5 to 10 minute fictional narrative, documentary, or news broadcast. In Project 3-1 you completed a script breakdown on the script you wrote for Project 2-2, with each scene on its own scene breakdown sheet.

When we look at scheduling a shoot it probably is going to be the most difficult to schedule a fictional narrative, especially if the script calls for a number of actors and locations. If you are shooting a news broadcast, for example, you probably shoot in a specific location at a specific time.

Scheduling a documentary can be tricky, but most of what is to be scheduled is driven by factors beyond the control of the documentary maker. If, for example, the documentary is based on a scheduled event, taping has to be done while the event is taking place. If you are working with a specific person, you have to be where they are whenever they are willing to let you come along. That does not mean you do not have to schedule, but you have to schedule around the other people or events.

As with the script breakdown, the fictional narrative that you created in Lesson 2 is the basis for this section on strip boarding, scheduling, and budgeting.

Create a Strip Board Using Microsoft Excel

Many professional film and video productions use what is called a strip board. The strip board consists of two parts: (1) the strip board that identifies the information and holds the strips, and (2) the strips themselves. Each strip has the elements needed to shoot each scene. The strip for each scene is placed on the strip board according to when and where each scene will be shot. When a scene has been shot, the strip for the scene is pulled off of the strip board so you know the scene is finished. The purpose of a strip board is to put all of the information for what needs to be shot in one location instead of having to shuffle through pages and pages of script or scene breakdown sheets. A quick glance tells you everything you need to know about when to schedule people, locations, and equipment. You also can go back to the individual scene breakdown sheets if you are unsure of something.

STEP-BY-STEP 4.1

In this Step-by-Step activity, you create a strip board using Microsoft Excel and fill in the basic information you need. The figures in this Step-by-Step are taken from a MAC.

1. Open Microsoft Excel.

2. If the Project Gallery comes up, click **Excel Workbook** from the options, then click **Open**. A new, blank document appears.

3. Click **File**, then click **Page Setup** from the menubar. Click the circle next to **Landscape**. This turns the paper on its side. This gives you more area across the page so you can view more scenes. Click **OK**.

> **Note**
>
> If the Project Gallery does not appear, and a new workbook does not appear either, you can create a new workbook by pressing **Ctrl+N** on a PC, or **Command+N** on a Mac.

4. Format the first column by moving the cursor to the line between columns A and B. Click and drag the line to the right until the width is about 18.00. (See Figure 4-1.)

STEP-BY-STEP 4.1 Continued

FIGURE 4-1
Adjusting column width

Move cursor to the line between columns A and B

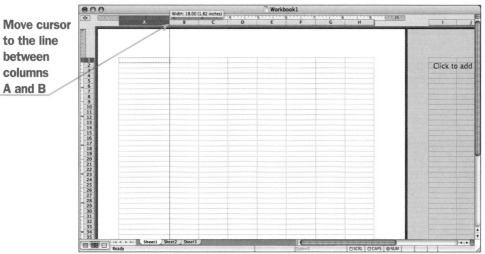

5. Place the mouse on the line between the 7 and the 8 on the left side of the window. Click and drag down to increase the height to 93.00. (See Figure 4-2.)

FIGURE 4-2
Adjusting row height

Place the cursor on the line between the 7 and 8 on the left-hand side of the window

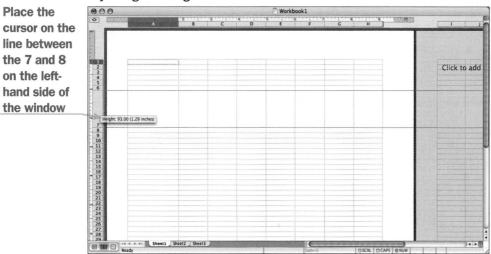

6. Click the letter **B** at the top of the workbook window and drag to the right until you reach the P column. Release the mouse. Columns B through P are highlighted.

7. Click **Format**, point to **Column**, and then click **Width** from the menubar. The Column Width dialog box appears.

STEP-BY-STEP 4.1 Continued

8. If you are using a PC, type **5** in the column width box. Click **OK**. If you are using a Mac, type **.5**. This makes columns B through P a uniform width. Figure 4-3 shows the basic strip board layout. Column A is where you identify the separate pieces of information and columns B through P is what the strips look like.

FIGURE 4-3
Basic strip board layout

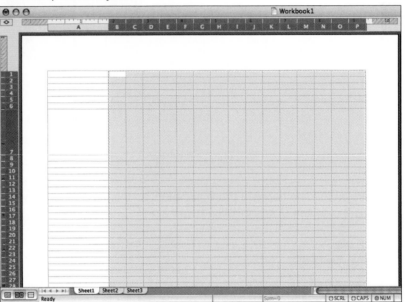

9. Place the cursor in the first cell, A1.

10. Type **Sheet Number** and press **Return** or **Enter** to move to the next cell below.

11. Type **Scene Number** and press **Return** or **Enter** to move to the next cell below.

12. Type **DAY or NIGHT** in cell A3, type **INT or EXT** in cell A4, type **Page Count** in Cell A5, type **Shooting Day** in Cell A6, and type **Character** in A7. Your workbook should look like Figure 4-4.

STEP-BY-STEP 4.1 Continued

FIGURE 4-4
Filling in Information from breakdown sheet

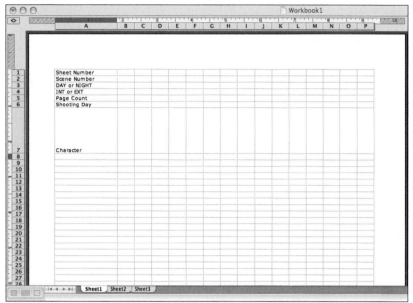

13. Type the numbers **1** through **10** in Cells A8 through A17. Type **1** in cell A8, then press **Return** or **Enter**. Type **2** in A9, then press **Return** or **Enter**. Type **3** in A10, then press **Return** or **Enter**, and so forth until you type **10** in cell A17. Then press **Return** or **Enter**.

14. Type **Extras** in cell A18. Press **Return or Enter**.

15. Click in cell A8 and drag the mouse down through cell A17 to select those cells.

16. Click **Format**, then click **Cells** from the menubar.

STEP-BY-STEP 4.1 Continued

17. Click the **Alignment** tab, then click the drop-down arrow next to General under Horizontal. Click **Left (Indent)**. Click **OK** and click anywhere to deselect the cells. Your screen should look like Figure 4-5.

FIGURE 4-5
Adding information to the script layout

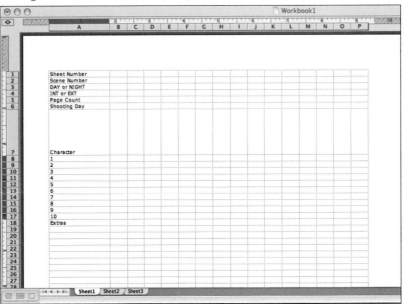

18. Click **File**, then **Save** from the menubar.

19. Save as **SBS4-1.xls**. Press **Return** or **Enter**. Leave the file open for the next Step-by-Step activity.

Now that you have the basic layout you need to get it ready to print.

STEP-BY-STEP 4.2

In this Step-by-Step activity, you create borders around the cell, define the area to be printed, and then print the storyboard.

1. With **SBS4-1.xls** open, select cells **A1** through **O18** by clicking in A1 and dragging to cell A18, and then dragging to the right to Cell O18.

2. Click **Format**, then click **Cells** on the menubar. The Format Cells dialog box appears.

3. Click the **Border** tab from the tabs along the top of the window.

4. Click the square buttons labeled **Inside** and **Outline** and click **OK**. You now have black borders around each cell as shown in Figure 4-6.

STEP-BY-STEP 4.2 Continued

FIGURE 4-6
Adding cell borders

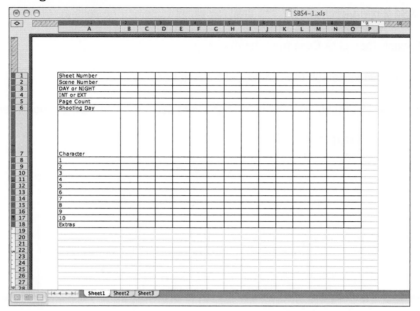

5. While all of the cells you selected earlier are still selected, click **File**, point to **Print Area**, and then click **Set Print Area** from the menubar. This creates a dashed line around the outside of the cells you want to print.

6. Click **File**, then click **Print** from the menubar.

7. Use the arrows next to the **Copies:** or **Number of copies:** field (depending on the version of Excel that you are using) to change the number of copies to 2. Click **OK**. You now have two copies of your strip board.

8. Get a piece of cardboard and glue one copy of the strip board to it. You want the strip board itself to be a board.

9. Using scissors or a paper cutter, cut the second copy along the horizontal lines. You can throw the far left column with the text on it away. You should now have about 12 long, skinny pieces, or strips.

> **Did You Know?**
>
> You always can check to see what is going to print by clicking **File**, then clicking **Print Preview** from the menubar.

10. Save your strip board as **SBS4-2** and leave the file open for the next Step-by-Step activity.

If you choose to fill the strip board out on the computer before you print it out, you have one more step to do.

STEP-BY-STEP 4.3

In this Step-by-Step activity, you format the strip board if you decide to fill it in electronically.

1. With **SBS4-2** open, click cell **B7** and drag to cell **P7** to select the cells in row 7.

2. Click **Format**, then click **Cells** from the menubar

3. Click the **Alignment** tab on the Format Cells dialog box.

4. Type **90** in the Orientation window in the right-hand section of the dialogue box. (See Figure 4-7.)

FIGURE 4-7
Changing the text orientation

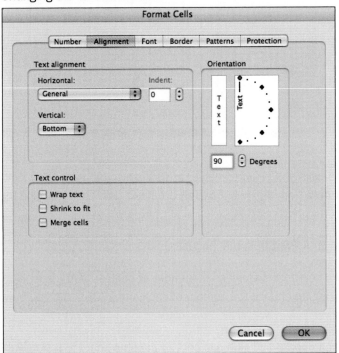

5. Click **OK**. Any text you type into any of the cells in the seventh row will be oriented sideways. When you type in the information, the text has a normal alignment, but when you press **Return**, the text displays sideways.

6. Type the name of each character in the script in the cells numbered 1 through 10. In cell A8, type **Jake**, then press **Return** or **Enter**. In cell A9, type **Pops**, then press **Return** or **Enter**. In cell A10, type **Mom**, then press **Return** or **Enter**. (See Figure 4-8.)

STEP-BY-STEP 4.3 Continued

FIGURE 4-8
Adding character names

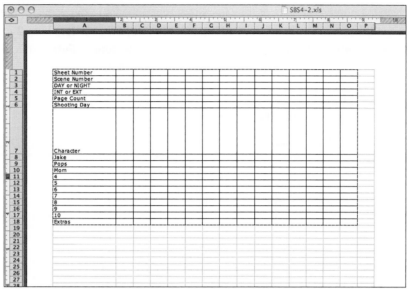

7. Print two copies so the strips and strip board have the same number of rows.

8. Save the document as **SBS4-3**. Leave the document opened for the next Step-by-Step activity. The strip board is ready.

S TEP-BY-STEP 4.4

In this Step-by-Step activity, you transfer the information from the breakdown sheet to the strip board. You can fill in the information by hand or electronically.

1. With **SBS4-3** and **SBS3-4** open, write or type **1** in the B1 cell of the strip board. This tells you that the scene is on breakdown sheet number 1. Press **Return** or **Enter**.

2. Write or type **1** in the B2 cell of the strip board. This is the scene number. Press **Return** or **Enter**.

3. Write or type **D** in the B3 cell of the strip board. This tells you the scene in shot during the day. Press **Return** or **Enter**.

4. Write or type **I** in the B4 cell of the strip board. This is because the scene in shot inside. Press **Return** or **Enter**.

5. Write or type **4/8** in the B5 cell of the strip board. This is the number of script pages the scene takes. If 8-Apr appears, format the cell as a fraction by clicking **Format** and then clicking **Cells**. If necessary, click the **Number tab**. Under Category, click **Fraction**, and then click **As eighths (4/8)** under Type:. If 4/8 does not appear in the cell, retype **4/8** in cell **B5**. Press **Return** or **Enter**.

STEP-BY-STEP 4.4 Continued

6. Write or type **Kitchen** in the B7 cell of the strip board.

7. Write or type the number of each of the characters that appears in the scene in the appropriate box. Type **1** in the B8 cell below the kitchen cell, then press **Return** or **Enter**. Type **2** in the B9 cell, then press **Return** or **Enter**. Type **3** in the B10 cell, then press **Return** or **Enter**. This tells you that Jake, Pops, and Mom all appear in this scene as shown in Figure 4-9.

FIGURE 4-9
Adding the number of each character

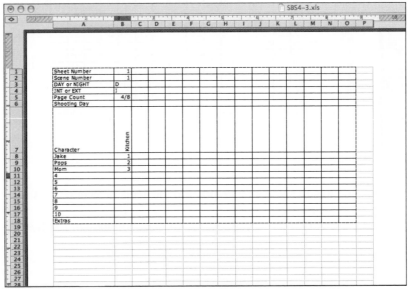

8. Use double-sided tape or doubled-over tape to tape the first strip into place in the second column from the left. See Figure 4-10 for an example of a strip board.

9. Save the document as **SBS4-4**. Close the document.

STEP-BY-STEP 4.4 Continued

FIGURE 4-10
Example of a strip board

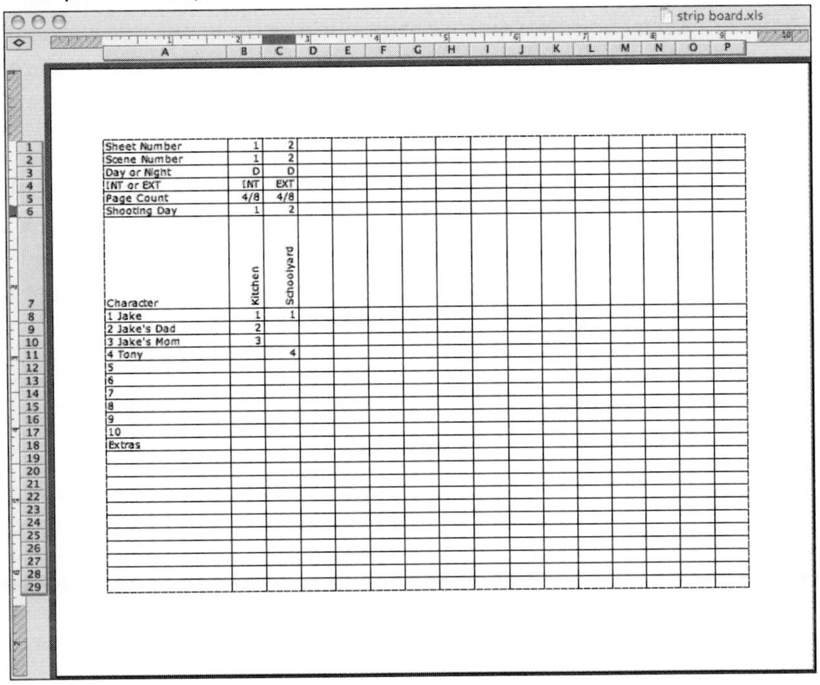

Sheet Number	1	2	
Scene Number	1	2	
Day or Night	D	D	
INT or EXT	INT	EXT	
Page Count	4/8	4/8	
Shooting Day	1	2	
Character	Kitchen	Schoolyard	
1 Jake	1	1	
2 Jake's Dad	2		
3 Jake's Mom	3		
4 Tony		4	
5			
6			
7			
8			
9			
10			
Extras			

You now have a quick, visual reference for what you need to have to shoot scene one of the script. As you complete a strip for each scene, organize the strips according to location. All of the scenes shot in the kitchen would be together, while all of the scenes shot at the school would be together. The scenes may not be in order, but you now know which scenes can be shot at each location.

Casting Each Role in the Script

When big name producers or big name directors want to make a movie they call big name movie stars and tell them they want to make a movie. When regular actors want a job, however, they have to try out for the part in order to be cast in the role. A try-out for an acting job is called a casting call. The process of casting for a movie goes something like this: The producer and director hire a casting director. A casting director works with talent agencies and the talent agencies know actors that would fit in any variety of roles. The casting director always is looking for new talent and is familiar with experienced actors. Because he or she knows many actors, a casting director can help the director identify actors that would fit well in any role. The casting director gives the director headshots (portraits) of the actors.

The director identifies, from the headshots, the actors he or she would like to have audition for each role. The casting director schedules the casting call in a location where actors can audition in one room, while the other actors wait their turn in another room.

The actor auditions for the director and casting director, but the producer may be there, as well as whomever else the producer or director feels needs to be there.

The actors may be asked to do a monologue or improvise, but most likely are given a scene from the script they are auditioning for. The actors do a cold reading of the script. A cold reading means they have not seen the script before. Somebody else in the room, such as an assistant, reads with the actors so the director can get a better feel for how the actors react and interact with the script. The director may give the actors directions or ask for a reading in a different way, and the actors need to be ready for those possibilities.

After all of the actors have auditioned they go home and wait for a callback. A callback means that the actors are seriously being considered for a role and the director wants them to come back and audition again. It is similar to a second interview for a job.

At callbacks the actor usually is asked to read again, but this time he or she may be asked to read with another actor being considered for another role in the same script. Most likely, the director most interacts more with the actor than he or she did in the first audition.

The producer or director then makes a decision and has the casting director offer the part to whomever he or she feels will do the best job.

You already may have people you want to act in your video, or know which people you want to be the anchors for your newscast. Use them, if they are willing, but remember that casting is an important element of any video production. It is important that each character is played by an actor who fits the role.

STEP-BY-STEP 4.5

In this Step-by-Step activity, you create the cast for your script.

1. Open **SBS2-8.fdr**. Read the script and create a mental image of each character. The script should give a general age range, if it is necessary to the character. If the character is supposed to be a grandfather, either cast somebody that is old enough to be a grandfather or have a really good make-up artist. Imagine what the characters look like and how they act. Are they mean? Are they overly-nice? Are they intense? Are they not very bright?

2. In a word processing program, write a quick physical description of each character. Include things like scars or black hair, but remember you can create scars with make-up or dye an actor's hair black. Some physical attributes, like height or muscle structure, are harder to find. People with these attributes are who you most likely have to go look for.

3. Hold an open casting call. Put the number of actors you are looking for, including how many males or females, on bulletin boards or in the school paper. Something like this probably would work:

Open Casting Call for short video project

Monday, April 5, 3:30. Rm. 302

Two males and One female needed

4. Save the file as **SBS4-5**. Close the file and program.

That is really all you need to say. Now just show up at the right place at the right time and see who shows up. If you are casting for a news broadcast and need an anchor for your project, make sure the candidate pronounces words clearly and "looks like" an anchor. Watch the nightly news and how the anchors deliver the news. Anchors look and sound like the authority on the day's events. They appear confident and deliver the news, whatever it is, with authority.

Know What to Look for When Scouting for Locations

Once you have broken down the script and know what locations you need, you have to go and find those locations. You need to find a place that looks like what is written into the script. You already may have a few locations picked out. It may be your own home, a friend's house, or wherever. There are a few things you need to think about, however, before you make final plans.

One of the fun things about making videos is that you can make almost any location look like somewhere else. Be creative. Many movies were supposed to take place in California or New York, but were actually shot in places like Utah and Toronto, Canada. Be creative and have fun with your locations.

First, make sure you have permission to be there. Whether you want to film in a school, a restaurant, or somebody else's house, you need to have permission (sometimes even written permission) to be there. You certainly do not want to drag cast and crew to your school on a weekend and get arrested for trespassing.

Gaining permission is usually pretty simple—just ask. If you do not get a favorable response to your request, thank the respondent for his or her time and look for another place. It will not do you any good to argue, and there are plenty of other places that will work. Check with the principle or manager, or whoever is in charge.

Second, make sure you have power outlets or electricity. I have shot in rooms without scouting first and found out there was not enough power for my lights and camera equipment. If you have enough good light to shoot without lights, all the better, but make sure you know what the light looks like around the time you plan on shooting. Check to make sure you have enough outlets close to where you are shooting. Also, make sure that you have enough outlets on different circuit breakers or you could blow a fuse at the wrong time. If the outlets are far away you can use extension cords, but you have to know you need them. You also can take care of the power issue with batteries or generators. If you do not prepare to use batteries, however, you may not bring them along. If you use generators you need to put them in an area that will not ruin the audio.

The third thing you need to consider is the sound. It is very frustrating to head out to the perfect location, set up the equipment, and then get bad audio because of passing cars, construction cranes, or a barking dog. Often an audience is more forgiving of bad video than bad sound. When scouting for a location make sure you know what is going on around you and what you need to do to get good audio. If you find a good location, but the neighbors have a dog that likes to bark at other dogs down the street, ask if you could have the neighbors take the dog into their house while you shoot. Again, most people are willing to help out if you ask.

Fourth, consider transportation. Just the other day I drove quite a while to pick up a piece of equipment for a shoot, only to find that the last piece did not fit into the car. It was just too big. I did not plan ahead. Make sure you plan well ahead so that cast, crew, and equipment get to where they need to be. How is everybody going to get where you are going? Is the place easy to get to? Does everybody have a map? Are there places to park? Find out all of this information long before you try to shoot so everyone can be there.

Fifth, consider food and water. They say an army moves on its stomach. Well, a film crew is an army. If you are shooting in a location where there is not much water, or it is not very convenient to get food or water, it is important to make sure the cast and crew are taken care of. You do not need much for a small volunteer cast and crew, but water and a few snacks are a good idea, especially in a remote location.

STEP-BY-STEP 4.6

In this Step-by-Step activity, you locate a place to shoot the script you broke down in Lesson 3.

1. Open and print out **SBS3-4.**

2. Open Microsoft Word or another word processing program of your choice.

3. Write down the answers to the following questions based on your breakdown sheets.

   a. Where are you shooting the scene? Give a detailed description of the location.

   b. Did you obtain permissions? From whom?

   c. Are there enough power outlets and where are they?

   d. Are there any surrounding sounds that are distracting and may interfere with the audio?

   e. Consider transportation issues. How is everyone going to get to the shoot location? Is the place easy to get to? Does everyone have a map? Are there places to park?

   f. How much food do you need and who is responsible for bringing it?

4. Save your file as **SBS4-6** and close the application.

Know What You Need for a Crew

As mentioned earlier, film and video are collaborative arts. You are going to need all the help you can get to make a video, especially if you have a large cast.

While the list of what you could have on a crew is pretty big, you want to make sure you have enough people to do the work required. On a small crew people can perform more than one task. For example, you can direct and run a camera, as well as set lights and move equipment between shots. You also can monitor audio during the shoot.

If you have people that are willing to help, such as classmates or friends and family, a basic crew might include the following:

Director. The director is in charge of the artistic side of things. He or she makes the final say of what is seen and heard in the final production. He or she determines what shots are needed and has the final say on how they look.

Director of Photography. In movies you have a director of photography, or DP. The director tells the DP how the scene should look and feel, and the DP makes it happen through the composition of the shot and the lighting.

Camera Operators. On a multi-camera shoot, you have camera operators who are running the camera and moving and adjusting the shot according to the director of photography's instructions, but they are camera operators, not directors of photography.

Audio Operator. This is the person who sets up the audio connections and makes sure the audio works the way it should. The audio operator also monitors the audio as it is being recorded to make sure sounds are recorded clearly and without glitches. I would suggest at least one other person to handle the audio. Sometimes it is difficult to operate the camera and monitor audio at the same time.

Gaffer. The gaffer is the electrician. He or she makes sure the equipment has the necessary power and that the equipment is where it needs to be.

Grip. A grip is a person who moves objects around and puts them where they need to be. The equipment may include the camera, lights, a dolly, cranes, etc.

While this is a basic crew you also may want someone to do the make up and take care of the craft services, which is food, snacks, and drinks for the cast and crew. Most film and video shoots require a lot of time spent on a set without time for the cast and crew to get something to eat and drink. Craft services help keep everybody happy and full of energy.

S TEP-BY-STEP 4.7

In this Step-by-Step activity, you identify the names of the crew to shoot your script.

1. Open **SBS2-8.fdr**.

2. Identify a crew to shoot your script. If you have a multi-camera shoot you want to have enough camera operators to run the cameras you need. Talk to the people you want on your crew and give them an idea of when you are going to shoot and how long you need them.

3. In a word processing program, type the names of your crew and indicate their position: Director, Director of Photography, Camera Operators, Audio Operators, Gaffer, Grip, and Craft services.

4. Save the file as **SBS4-7** and close the program.

Scheduling a Fictional Narrative

Now you have a strip board that tells you which scenes should be shot together, as well as who needs to be there. How do you actually use this information to schedule? I know you still have questions, but probably the most important question you have is, "How do I know how long it is going to take to shoot each scene?" The answer is, "I don't know."

How long it takes to shoot a scene depends on the complexity of the scene and how many camera setups it requires. A camera setup occurs is any time you move a camera. For a typical conversation scene with two characters you can plan on at least three setups. First you set the camera up for the establishing shot. The establishing shot gives the audience an understanding of where everyone and everything is in relation to each other. You shoot the entire scene from the establishing shot. You then re-set the camera for a close-up of the first character and shoot the entire scene from that setup. You would then re-set to a close-up of the second character and shoot the entire scene from that setup. It may sound time consuming, but it is surprising how quickly you can shoot a scene like this. With experience you may be able to only shoot what you know you want from each setup instead of shooting the entire scene in each one.

A more complex scene, like a carefully choreographed fight, could take days to shoot because of all of the setups and stunts. Even though a scene may take multiple days to shoot, it only may appear on screen for a minute or so. The first time I was on a set it took a full day to shoot. That whole day's worth of work was only on screen for about 30 seconds.

Some people say you can average two pages a day, while others say you can shoot as many as ten in a day. It all depends on what you are doing. You have to read your script and figure out how many different setups you need. Do not, however, do what I did on my first shoot and set up for each and every scene. I started at the first position, then moved to the second position, then back to the first position, then to a third position, then went back to the first position, then went to the third position, and then back to the second position . . . it took forever. With planning and some understanding of what you are doing you can keep the number of times you move the camera to a minimum.

Now look at your strip board. You need to first schedule any fixed dates. For example, you may be able to shoot at the school on Saturdays only. Schedule who and what you need to shoot at the school for Saturday. You may look at your strip board and decide that you have several scenes to shoot at the school and you need multiple days there. That is fine, just put it into your schedule.

Once you have determined which days you are going to shoot, write them down. Day 1 may be on Monday, while day 2 is on Thursday. Write which shooting day you shoot each scene on the strip for each scene. Scenes 1 and 3, for example, might be shot on day 1, while scenes 5 and 6 might be shot on day 2, and the school shots might be shot on Saturday, which is shooting day 3.

Next you need to schedule cast and crew. The time you want the crew there is called crew call. You want crew there as early as possible before you actually start shooting so you can set up camera and equipment. The cast, however, may not need to be there until the crew is set and ready to actually shoot. You need to determine how much time you need to set up, and then determine when you shoot. The cast also may need time to prepare make-up and wardrobe. I cannot tell you when to have cast and crew calls; that is determined by what you are shooting. I can tell you, however, that you do not want a lot of people standing around doing nothing.

Communication is the key. Make sure everyone involved knows where they need to be and when to be there. Make sure every scene has been scheduled. Use the scene breakdown sheet to make sure you have everything that you need to shoot each scene, and that the information is on the strip board so you easily can identify what the plan is for each day. Most importantly, write everything down. Use the breakdown sheet to write down when you are going to shoot the scene, and write down the crew and cast that need to be on set. Keep track of your information so you know what is going one and when it is going to happen.

STEP-BY-STEP 4.8

In this Step-by-Step activity, you use your strip board to schedule dates, crew, and cast for the scene with Jake and his Pops.

1. Open **SBS4-4.xls**.

2. In the Shooting Day row (B6), type **DAY 1**.

STEP-BY-STEP 4.8 Continued

3. Save the file as **SBS4-8.xls**. Close the file, but leave the application open for the next Step-by-Step activity.

4. Open **SBS3-4.doc** and type **Crew call is at 5:00 a.m.** and **Cast call is at 7:30 a.m.** in the production notes of the breakdown sheet.

5. Save as **SBS4-8.doc**. Close the application

Create a Budget Sheet Using Microsoft Excel

Now that you have an idea of cast and crew, where you will shoot, and an idea of how long you will shoot, you can create a budget. The budget needs to be flexible, as it will change.

Many major motion pictures have gigantic budgets. One hundred million dollars seems to be a small budget these days. You can make a video for much less. Even though you do not have a big budget, you still need to make a budget.

What kind of things do you need to budget for? For a big movie you need to budget transportation, cast, crew, insurance, film processing, script writing, script re-writes, the art department, and on and on and on. But do not worry about all of that right now. Focus on just the basics of tape stock, equipment rental, craft services (food and snacks to the uninitiated), transportation, props, wardrobe, and make-up. I am going with the assumption that the cast and crew are volunteers, but if they are not you have to budget for them as well.

STEP-BY-STEP 4.9

In this Step-by-Step activity, you create a budget sheet using Microsoft Excel. You start out with a budget of $400.00. The figures in this activity are taken on a MAC.

1. Open Microsoft Excel and start with a blank worksheet.

STEP-BY-STEP 4.9 Continued

2. Type **Tape Stock** in cell A1, then press **Tab**. Type **Equipment** in cell B1, then press **Tab**. Type **Food** in cell C1, then press **Tab**. Type **Transportation** in cell D1, then press **Tab**. Type **Props** in cell E1, then press **Tab**. Type **Wardrobe** in cell F1, then press **Tab**. Type **Makeup** in cell G1, then press **Tab**. You may need to adjust the column width. See Figure 4-11.

FIGURE 4-11
Adding scene breakdown elements

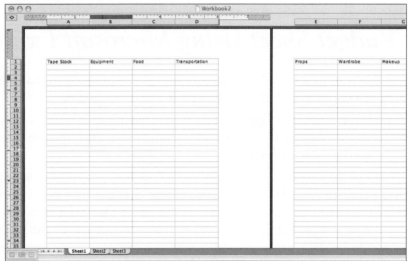

3. Type **75** in cell A2, then press **Tab**. Type **125** in B2, then press **Tab**. Type **25** in C2, then press **Tab**. Type **100** in D2, then press **Tab**. Type **25** in E2, then press **Tab**. Type **25** in F2, then press **Tab**. Type **25** in G2, then press **Tab**.

4. Click cell **A15**. Click **Insert** and then click **Function** from the menubar. A function is a mathematical formula that makes Excel perform all of the number crunching for you.

5. Click **Math&Trig** from the function category list, and use the scroll bar for the function name window to find and click **SUM** as shown in Figure 4-12. Click **OK**.

STEP-BY-STEP 4.9 Continued

FIGURE 4-12
Adding the columns

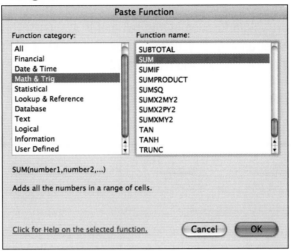

6. A2:A14 should appear in the Function Argument window next to the Number 1 field as shown in Figure 4-13. If it doesn't, type A2:A14. This totals any numbers you enter into cells A2 to A14. Press **OK**. The number that appears in the cell should match the number in Cell A2. If it doesn't, make sure that the formula is A2:A14.

FIGURE 4-13
Function Argument window

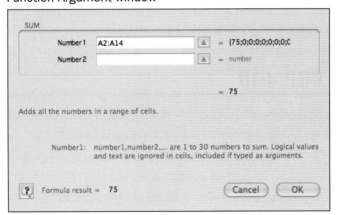

7. Click cell **B15**. Click **Insert** and then click **Function** from the menubar.

8. This time the function category window should display **Most Recently Used** at the top of the window. If not, then select it from the drop-down list. Click **SUM** if necessary and then click **OK**.

9. In the Function Argument window next to the Number 1 field make sure it says B2:B14. If it doesn't, type **B2:B14** into the window. Click **OK**.

STEP-BY-STEP 4.9 Continued

10. Insert the same formula in each cell of Row 15 until you reach cell G15. The formula for C15 should be C2:C14, the formula for D15 should be D2:D14, and so on.

11. Click cell **H15**. Insert a Sum function in this cell that is A15:G15. This will give you a total budget number so you can get an idea of your total budget of $400. It will appear in this formula. (See Figure 4-14.) Set all of the columns widths to **14.75**.

FIGURE 4-14
Insert the sum of the total budget

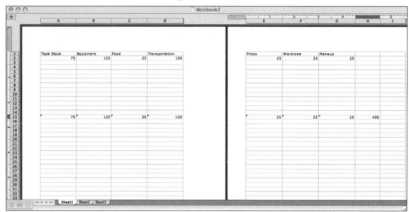

12. Save the document as **SBS4-9**. Leave it open for the next Step-by-Step exercise.

You are using a Sum function to do your math, which means that any time you enter money spent into the formula it needs to be a negative number.

STEP-BY-STEP 4.10

In this Step-by-Step activity, you move money from one budget item (tape stock) to another (wardrobe). This demonstrates how to both add to and subtract from your budget

1. With **SBS4-9** open, click cell **A3**.

2. Type **-24.50**. Press **Return** or **Enter**. The total for column A in cell A15 will change to 50.5, and the number in H15 will change to 375.5.

3. Click cell **F3**.

4. Type **24.50**. Press **Return** or **Enter**. The number in cell F15 will change to 49.5, and the number in H15 will change back to 400. (See Figure 4-15.)

STEP-BY-STEP 4.10 Continued

FIGURE 4-15
Adding and subtracting from the budget

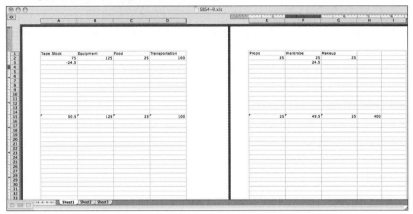

5. Save as **SBS4-10**. Close the document. Exit Microsoft Excel.

Know How to Storyboard

Storyboarding is one of the most enjoyable parts of the office work for me. The **storyboard** is a visual guide for the shoot. The strip board tells you who needs to be where and when and what will be shot, but the storyboard helps you see what each shot looks like. Many times when I am working on a script I storyboard the show out as I write so I can keep everything that is going on clear in my mind.

Go through the script and visualize each scene. What is the camera seeing? What is the composition? What does the audience need to see? Simple drawings will work. If you want a large, more expensive production, you can hire a storyboard artist. You can ask a friend to draw the storyboards for you if you wish.

The storyboard should appear like a comic book, with the page divided into maybe six squares. The squares are frames for each drawing and should be big enough to give an idea of what is seen in the shot.

Storyboard each frame until you have a good idea of what the scene looks like. For example, for a scene involving two people having a conversation, the first frame should show the establishing shot. The lines that correspond to the shot should be written underneath. You do not have to have each and every line for the shot, but enough to get an idea for how much of the dialog is covered. If the scene then goes to a close-up, draw how that close-up looks, with a bit of dialog under it to give an idea of how much dialog that shot covers. The next frame might then be a close-up of the second actor, or of the establishing shot.

For more complex scene, say a car chase, the storyboard should give an idea of what each shot should look like. Remember, this is a guide for the director and camera operator more than for the actors. While a conversation scene has relatively few storyboard frames, a more complex scene may have a great number of storyboard frames to show what is happening.

Storyboards should not be polished works of art. They need to be done quickly, but they need to be accurate and clear enough that a quick glance tells the director, DP, or anybody else what the shot looks like. Concentrate on the composition for each camera setup. For example, if the shot shows three people, where are they in relationship to each other? Where are they in relationship to the camera?

Camera moves, such as a pan, can be shown with an arrow pointing in the direction the camera moves and a quick explanation of the move, like "Camera pans right" written under the frame.

STEP-BY-STEP 4.11

In this Step-by-Step activity, you create a storyboard for your script.

1. Open **SBS2-8.fdr**.

2. Take a piece of blank white paper and divide it into six blocks.

3. In the first paragraph of the script, Jake is standing next to the sink reading a magazine. Draw in the first block what you read in the first paragraph.

4. In the second block, draw the second paragraph where Pops walks into the room dressed in a conservative suit looking much younger than his 54 years, and puts his briefcase on the table.

5. In the third block, draw an outscreen bubble of Mom saying "Dallas, you are not going to wear that suit again. It's embarrassing."

6. If you created your document electronically, save it as **SBS4-11** and close the application. If you created it by hand, submit the document to your instructor.

SUMMARY

In this lesson you learned:

- That a strip board is a visual guide to help you identify what will be shot and who needs to be involved each day of a shoot. You also learned how to create a strip board using Microsoft Excel.

- What kinds of things need to be scheduled in order to prepare for a shoot, including cast, crew, equipment, and locations.

- How to create a budget sheet using Microsoft Excel.

- How to create a budget for a small shoot and the kind of items that need to be included in such a budget, including cast, crew, equipment, and possible location and transportation costs.

- How to identify cast members that will best fill each role in a script.

- What the different crew positions are and how to identify what you will need for a shoot.

- What to be aware of when identifying a location for shooting, including accessibility needs, power needs, lighting and equipment needs, and how to identify possible audio problems.

- The purpose of a storyboard and how to create a storyboard that will help you in each shot.

VOCABULARY *Review*

Define the following terms:

audio operator	craft services	grip
camera operator	director	storyboard
casting	director of photography	strip board
casting call	gaffer	

REVIEW *Questions*

MULTIPLE CHOICE

Refer to the **strip board for multiple choice.pdf** and select the best response to the following.

1. According to the strip board, which character is in every scene?
   A. Jake
   B. Jake's dad
   C. Jake's mom
   D. Tony

2. According to the strip boards, which characters are in scene 1?
   A. Jake, Tony, Jake's mom
   B. Jake's mom, Jake's dad
   C. Jake
   D. Jake, Jake's dad, Jake's mom

3. According to the strip board, the location for scene 1 is
   A. Jake's house.
   B. the kitchen.
   C. the schoolyard.
   D. Tony's house.

4. On which shooting day will scene 3 be shot?
   A. day 1
   B. day 2
   C. day 3
   D. day 4

5. Which scene number is shot in the schoolyard?
   A. scene 1
   B. scene 2
   C. scene 3
   D. scene 4

6. Which day does the actor who plays Chrystal need to be on the set?
   A. day 1
   B. day 2
   C. day 3
   D. day 4

7. Which scenes are shot outside?
   A. scene 1 only
   B. scenes 2 and 3
   C. scenes 1 and 4
   D. scene 2 only

8. According to the strip board, how many days will this shoot take?
   A. 1 day
   B. 2 days
   C. 3 days
   D. 4 days

SHORT ANSWER

Write a brief answer to the following questions.

1. Explain the purpose of a strip board.

2. Describe what happens at a typical casting call and callback.

3. Explain what types of things you should look for and identify when scouting locations.

4. Referring to the strip board used to answer the multiple choice questions, why shoot scene scenes 1 and 3 on the same day?

TRUE/FALSE

Circle T if the statement is true or F if the statement is false.

T F 1. The strip board is used to help organize the shooting schedule.

T F 2. The storyboard must be a work of art.

T F 3. A setup refers to each time the camera is moved and placed for another shot.

T F 4. You can plan on shooting no more than a single page each day of a shoot.

T F 5. The gaffer is the head electrician.

T F 6. A casting call is a job interview for an actor.

T F 7. Functions are mathematical formulas used in Microsoft Excel.

T F 8. The grip is responsible for operating the camera on a multi-camera shoot.

PROJECTS

PROJECT 4-1

1. Create a strip board for Project 2-2 by using the scene breakdown sheets you created in Project 3-1. Save the file as **Project 4-1**.

PROJECT 4-2

In this project, you cast each role, locate a place to shoot, and identify a crew.

1. Open Project 2-2. Read the script and create a mental image of each character. The script should give a general age range, if it is necessary to the character. Imagine what the characters look like and how they act. Are they mean? Are they overly-nice? Are they intense? Are they not very bright?

2. In a word processing program, write a quick physical description of each character.

Put the number of actors you are looking for, including how many males and females.

3. In order to locate a place to shoot, write down the answers to the following questions based on the breakdown sheets you created for Project 3-1.
    A. Where are you shooting the scene? Give a detailed description of the location.
    B. Did you obtain permissions? From whom?
    C. Are there enough power outlets and where are they located?
    D. Are there any surrounding sounds that are distracting and may interfere with the audio?
    E. Consider transportation issues. How is everyone going to get to the shoot location? Is the place easy to get to? Does everyone have a map? Are there places to park?
    F. How much food do you need and who is responsible for bringing it?

4. Identify a crew to shoot your script. If you have a multi-camera shoot you want to have enough camera operators to run the cameras you need. Talk to the people you want on your crew and give them an idea of when you are going to shoot and how long you need them.

5. Type the names of your crew and indicate their position: director, camera operators, audio operators, gaffer, grip, and craft services.

6. Save the file as **Project 4-2**. Close the file and program.

PROJECT 4-3

Create a budget sheet for Project 2-2 using the budget items from Step-by-Step 4.9. If you already know how much money you have, plan your budget based on those numbers. If you don't have any money at this point, guess at what you think you might need. You will be surprised at how little money you actually need for a shoot if you go to second-hand stores (like the Salvation Army) for some of the items, like wardrobe and props. Save the file as **Project 4-3**.

PROJECT 4-4

Create a storyboard for Project 2-2. You can do this manually or electronically with a drawing program like Adobe Illustrator. Save the work as **Project 4-4**.

 ## WEB PROJECT

Go online and look for a movie script. Some have been produced, some have not. Some have a copyright, some do not. Be careful not to infringe on any copyrights because most produced scripts are for sale. You can buy one if you like, but the idea is to breakdown and do all the work you would need to schedule for at least four scenes from the script you find online. Fill out a strip board for the four scenes you have chosen.

TEAMWORK PROJECT

Team up with classmates, as well as family and friends and cast and crew, to shoot the script you wrote earlier. Film is a teamwork production, so learn how to make that happen. Assign people positions or roles in the video. Also be willing to help them as cast or crew for finishing their scripts.

CRITICAL *Thinking*

ACTIVITY 4-1

Watch a movie and break down the movie for cast and locations. Try to figure out how many different scenes were shot in a single location and how many days you think it would have taken to shoot those scenes. You could even go as far as to create a strip board for the script you think was shot. There are no right or wrong answers, but the more you pay attention to this project, the more you will recognize the work that goes into creating a film.

PRODUCTION

Unit 2

 Estimated Time for Unit: 6.5 hours

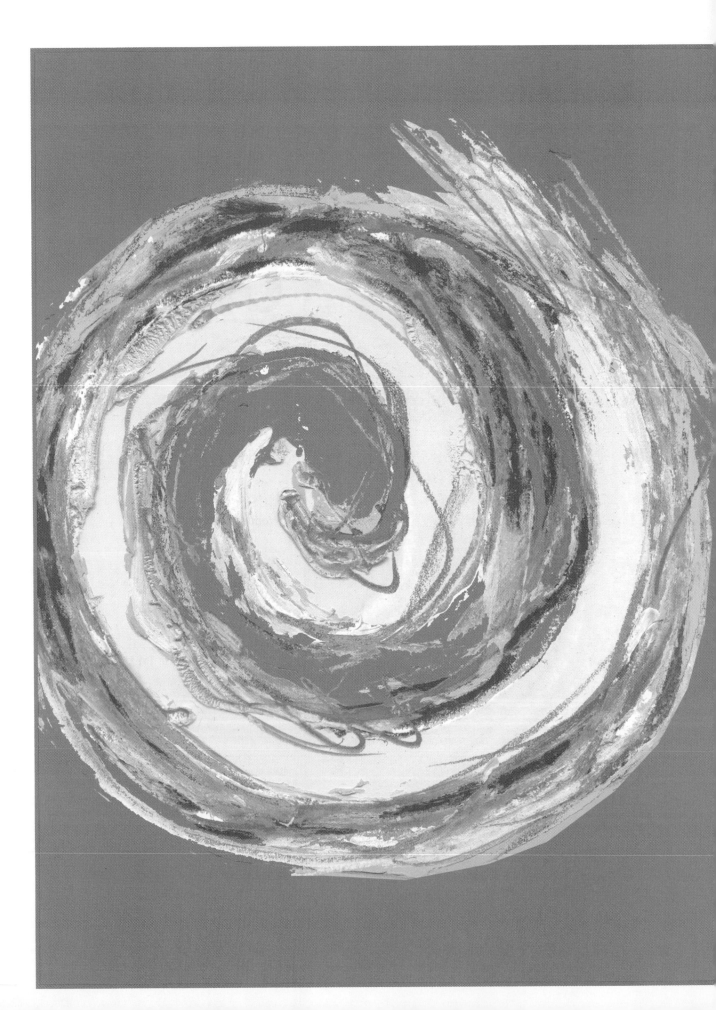

THE CAMERA

How Does the Camera Work?

The camera is the most important piece of equipment when it comes to video. Think about it – of all the expensive lights and spectacular video editing systems will not do you any good if you do not have a way to acquire video.

Most everything in the world reflects light. The light can come from the sun, the stars, a light bulb, a fire, or many other sources. Your eyes gather that reflected light and send signals about the light patterns to your brain. Your brain translates those signals into images that tell you that you are looking at a blue sky, a red rose, or a green car.

A video **camera** captures the light information and translates it into a video image in the same way. A camera **lens** gathers the reflected light and directs it to the brains of the camera that are light sensitive. There is an **imaging device in film camera** that translates the light into an image called film. The imaging device in a video camera is a **charge-coupled device (CCD)**. The electronic signal created by a CCD is recorded or stored on tapes, DVD's, or hard drives. See Figure 5-1.

FIGURE 5-1
Basic camera construction

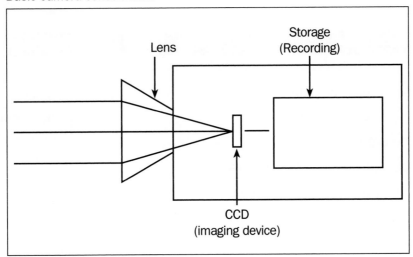

This is a very simple explanation of how a camera works, but it is enough to get started. Each camera has four distinct parts:

- The lens
- The imaging device
- Storage
- The viewfinder

How a Lens Works

The job of the lens on a camera is to direct light to the imaging device. Video lenses are **convex**, meaning that they are thin along the edges and wider in the middle. The light covers the surface of the lens when it enters, but the shape of the lens focuses the light to a single point, called the **focal point**, when it leaves the lens (see Figure 5-2).

Note

Some of what is covered here does not apply to consumer camcorders. Consumer camcorders are designed to be fool-proof, meaning that all a user has to do is point, shoot, and occasionally zoom in or out. While it is possible to shoot a nice image, some control, such as control over-exposure and depth of field, are taken from the camera operator and given to the camera. This is important information to understand if you plan on working in the video field professionally.

FIGURE 5-2
Example of focal point

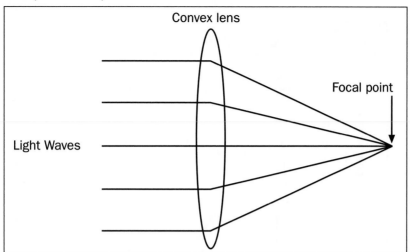

Changing the distance between the lens and the focal point changes the area that the lens can see. Photographic and video lenses are referred to according to their focal length. The basic idea of the focal length of a lens is the distance between the lens and the imaging device when the image is in focus (see Figure 5-3).

FIGURE 5-3
Focal length of a lens

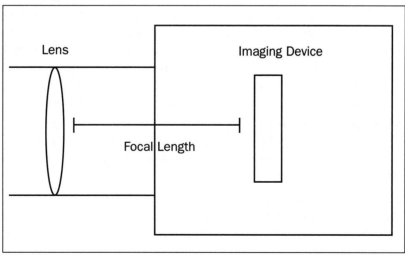

Don't worry about the technical definition. The important thing to know about focal length is what it means when you are using a video camera. The focal length is measured in millimeters, or mm. This means you see a lens referred to as a 50mm lens, a 400mm lens, a 1000mm lens, etc.

A lens with a shorter or smaller focal length has a wider field of vision, while a lens with a longer focal length has a narrower field of vision (see Figure 5-4).

FIGURE 5-4
Relation of focal length to field of vision

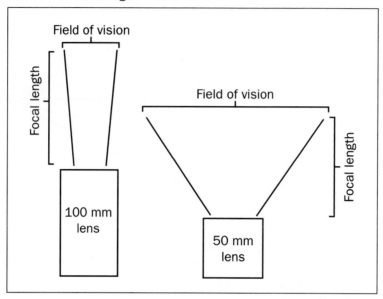

Professional cameras have removable lenses that allow the camera operator to determine which type of lens to use. The lens categories are normal, wide-angle, telephoto, and zoom.

The normal lens for a camera is a lens that sees what the human eye would see. The normal lens for most cameras is a 50mm lens. Anything under 50mm is referred to as a wide-angle lens, and anything over 50mm is a telephoto lens. A zoom lens is a lens with a variable focal length, meaning the lens can serve as all three of the other lenses. Many professional cameras use zoom lenses to make things easier. Zoom lenses are convenient, but they do have limitations and do not allow the camera operator as much control over the image as the separate lenses.

In a recent televised football game a player broke away for a 75 yard run. The camera at eye level in the end zone made it look like the defensive players could just reach out and touch the player with the ball. When they went to a camera along the sidelines, however, the defensive players were really 10 to 20 yards behind the player with the ball.

Why did the defensive players appear so close in the end zone camera when they were really so far away? The focal length compressed the distance between the players and made them appear to be closer than they actually were. Changing the focal length changes how characters look, how close they appear to be to each other, and can create a different mood in a scene.

A telephoto lens compresses the objects in the frame, meaning that it makes them appear closer together than they really are. A wide angle lens has the opposite effect, making objects appear further away than they really are.

Optical and Digital Zoom

One of the selling points for consumer camcorders is its optical and digital zoom capabilities. It is not uncommon to see lenses that offer a 30X optical and a 1200X digital zoom. But what does that mean? The numbers 30X and 1200X are ratios, which means that the lens can enlarge the image you see by 30 or 1200 times. Translation, the camera can get you a really good close-up from far away.

The optical and digital part of the equation is actually the important part. An optical zoom uses the lens elements to enlarge the image. The image should be as crisp and in focus using a zoom lens as it should be with a dedicated telephoto lens. The optical zoom is the "good" zoom. The only problem with such a high zoom ratio is that it is very difficult to hold the image steady without using a tripod or electronic image stabilization.

The digital zoom is the "bad" zoom. Digital zoom is accomplished by enlarging the pixels in the middle of the image. This means you have fewer and fewer pixels, but the individual pixels are bigger and bigger. Zooming in too close with a digital zoom could leave you with a lot of big pixels without knowing what the image really is.

About Exposure

Go outside with a mirror when the sun is out. Now look at your eyes in the mirror. How big are your pupils? Hurry inside and look at your eyes in the mirror again. How big are your pupils now? Are they bigger or smaller than when you were outside?

Your pupils automatically shrink down when there is plenty of light. This is how the eye controls the amount of light that reaches your brain. Your pupils open up (get bigger) when there is less light allowing more light to reach your brain.

The lens serves the same purpose for a camera. The lens controls the exposure, or the amount of light that reaches the imaging device. The part of the lens that determines how much light gets through is called the aperture. The aperture allows less light in when it is shut down and allows more in when it is open.

F-Stop

The size of the aperture's opening is measured in f-stops. Common f-stops are f/1.4, f/2, f/2.8, f/4, f/5.6, f/8, f/11, f/16, and f/22. Think of these f-stop numbers as division problems where f equals the focal length of the lens. If you have a 50mm lens, and you open the aperture to f/2, how big is the opening in the aperture? Take 50 (the focal length) and then divide by two and you have an aperture with a diameter of 25mm. If you set the aperture at 5.6 on the same lens, divide 50 by 5.6 and you get 8.928, meaning the aperture is roughly 9mm in diameter. Notice that a smaller f-stop number means a larger opening (see Figure 5-5).

FIGURE 5-5
Examples of common f-stops

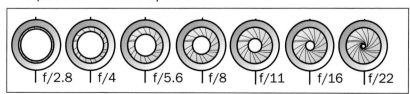

This also means that a longer lens has a larger aperture at the same f-stop setting. The aperture for a 100mm lens is 50mm in diameter at f/2, while the aperture was only 25mm for a 50mm lens at the same setting.

Another interesting thing about f-stops is that each setting allows half as much light in as the previous setting. This means that f/2 allows half as much light through as f/1.4, and f/16 allows twice as much light in as f/22.

How do you know the correct f-stop for a shot then? One way is to use a light meter. A light meter will tell you the correct f-stop for the amount of light you have. I highly recommend using a light meter.

A second way to tell if the exposure is right is to look through the viewfinder. If you can't see details because it is too bright, the shot is overexposed. Simply close down the aperture (move to a higher numbered f-stop, like to f/11 from f/2.8). The shot is underexposed if it is too dark to see the details. In this case open up the aperture.

Lens Care

Lenses are expensive, especially interchangeable lenses, so take care of them. Make sure to keep the lens cover on when the camera isn't in use. If you have a lens that will accept filters, use a clear filter when you shoot—it's less expensive to replace a scratched filter than it is to replace an entire lens. Also, use a lens hood if your lens will accept one. A lens hood attaches to the end of the lens and is like a baseball cap for a camera.

Check the owner's manual for the best way to clean the lens. If you don't have an owner's manual, however, hope is not lost. First, blow dust or anything else off, but don't use your mouth (you don't want spit on the lens). Second, use a lens cleaning cloth (you could use something else, but you risk scratching your lens) and start in the center of the lens. Make a gradually growing circle until you reach the outside edge of the lens. You also can use a lens cleaning solution, but put it on the lens cleaning cloth, not directly onto the lens.

Depth of Field

The f-stop not only controls exposure, but it helps determine the depth of field. Depth of field is the area of the image that is in clear focus. For example, if the subject is 12 feet away from the camera everything between 8 and 16 feet away from the camera might be in focus. A shallow depth of field means that there is only a small area around the subject that is in clear focus, for example, only objects that were 11 to 13 feet from the subject would be in clear focus. A deeper depth of field means that more of the area around the subject is in focus, for example, only objects that were 6 to 18 feet away from the subject would be in clear focus (Figure 5-6). The camera position, lens type, and f-stop all help determine the depth of field.

FIGURE 5-6
Depth of Field

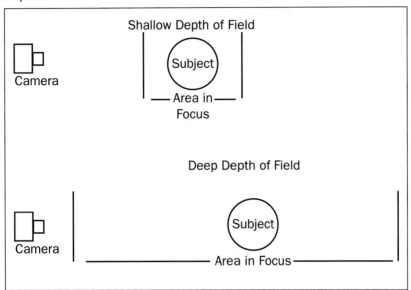

First, depth of field increases when you decrease the size (or close down) the aperture, and decreases when you increase the size of the aperture. If you shoot at f/22 you have a deeper, or greater, depth of field (meaning more of the image is in focus) than if you shoot at f/2.8.

Second, depth of field decreases when the subject is closer to the camera, and increases when the subject is further from the camera. You will have less depth of field when the subject is just a couple of feet from the camera than if the subject is 20 feet from the camera.

Third, decreasing the focal length of the lens increases the depth of field, while increasing the focal length decreases the depth of field. Using a telephoto lens will give you less depth of field, while using a wide-angle lens will give you more depth of field.

Charge-Coupled Devices

The lens focuses the light waves to a single point, which is basically where the **imaging device** is located. The imaging device is the part of the camera that translates the light waves into an image. The imaging device in a video camera is a charge-coupled device (CCD). The CCD is a light sensitive chip that changes the light waves into electronic images that are recorded onto tape, DVD, or some type of hard drive.

Most consumer cameras have a single CCD, or chip, that contains rows of pixels. Higher end (and more expensive) cameras have three separate CCDs. Each chip in a three CCD camera is sensitive to red, green, or blue light (you may recognize this as RGB). One chip, then, captures the red information, another captures the green information, and the third the blue information. Each CCD in a three CCD camera may capture as much information as the single CCD in a consumer camera. This means the image will be much more detailed with more contrast between light and dark areas.

Most people won't notice the difference between video from a consumer miniDV camera and video from a professional DV camera, but you notice it if you have ever watched one of those "send us home movies of you doing dumb things" television shows. The home videos are usually either under or overexposed, and the images are not as crisp and clear as the stuff shot in the studio.

How does each chip grab just the light that it is sensitive to? A three CCD camera has a series of prisms and mirrors that the light passes through before it reaches the CCDs. The prism splits the light and the mirrors direct the correct color information to the correct CCD (see Figure 5-7).

FIGURE 5-7
Prism directing the light waves to the correct CCD.

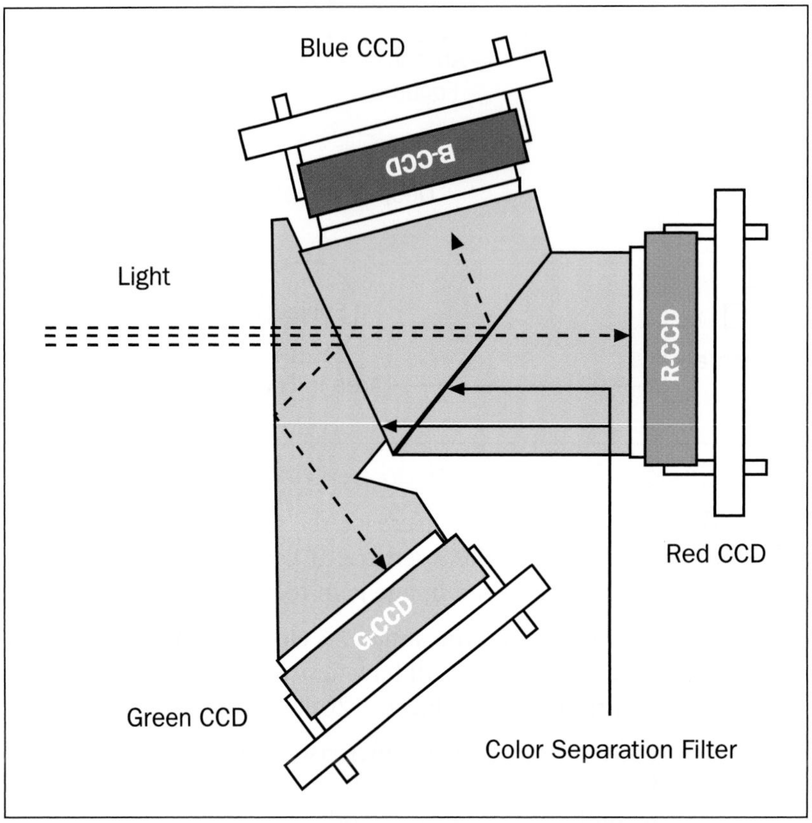

The Viewfinder

The next part of the camera that we will talk about is the viewfinder. The viewfinder allows you to see what the camera is shooting. Most consumer cameras these days have small LCD screens as well. One of the odd things (to most people) is that viewfinders on professional cameras are generally black and white, while the viewfinders and LCD screens on consumer cameras are in color. Why is this?

Think back to the discussion about luminance, hue, and chrominance in Lesson 1. The human eye is more sensitive to light and dark than it is to color information. Black and white allows the camera operator to obtain a clearer focus because he or she is able to see the contrast, where color viewfinders kind of blur the edges of the image. A consumer camera doesn't allow the user to focus manually.

Storage

The last part of the camera we will discuss is the storage. The storage for a film camera is the film. The storage for a video camera can be a tape, a hard drive, flash memory, or a disk. The most common storage method is tape and the most common tape is miniDV. The advantages of using tape are that it is cheap, easily accessible, and can be stored indefinitely.

Hard drive storage is becoming more and more common. One thing people have difficulty understanding is that the hard drive may store any different number of compression formats, depending on the camera and drives. The cameras I use for my day job use hard drives to store the video, but the format is DV. It is just another way to store information.

The advantage of hard drive storage is that you can transfer the video directly to the computer without having to capture the video. The hard drives in the cameras I use connect directly to my computer through a Firewire connection. The camera's hard drive appears as a hard drive icon on the computer's desktop and I simply copy the files to where I want them on my computer. Before firewire connections, it would take me three hours to capture three hours of video, however, now I can capture the video files (about 40 GB) in about 15 minutes. The down side of hard drive storage is that it is more difficult to archive. You need to transfer the video to another hard drive so you can reuse the hard drive in your camera.

Flash memory (often called solid state memory) is becoming more and more common. A regular hard drive has a lot of moving parts that can sometimes be knocked loose or out of joint if dropped. Solid state is like the USB drives you see everywhere, only bigger (and more expensive). The advantage of the solid state storage is that there are no moving parts to jar loose. The downside is the cost

One of the newer methods of storage is to record directly to a DVD. The advantages of using a DVD is that you can play it in a normal DVD player, and the DVD's are inexpensive. The downside of this type of storage is that the video is compressed, making it hard to edit and work with.

White Balance

Cameras are wonderful, but they aren't all that smart when it comes to knowing what white looks like. The problem is not really with the camera, but white light is not always white. The light waves we see come in different temperatures, referred to as color temperature. Some light has more blue in it, some has more red, some is a little greener, and so on. The next lesson will go over color temperature in depth.

When working with a camera it is important to help the camera know what is really white. Sunlight, for example, puts a bit of a blue tint on a white surface. By telling the camera that the white surface is actually white, however, it removes the blue tint from the image.

What happens if you don't white balance? Everything in the image will look blue, or red, or green, or whatever color the light gives off.

Most cameras have a white balance feature, even inexpensive consumer cameras. Consumer cameras generally take care of the white balance automatically, but it doesn't always work. The white balance feature is easy to access on professional cameras and is generally a simple flip of a switch at the right time. The white balance feature on consumer cameras, however, is buried in a menu that is a little harder to find. Refer to the user manual on how to access and use the white balance on the camera you are using.

S TEP-BY-STEP 5.1

In this Step-by-Step activity, you perform a white balance. The process for performing a white balance is the same no matter what camera you are using.

1. Place a white balance card or some other white item in the same light as the subject you are shooting. Do NOT white balance in one place and shoot in another. It won't work.

STEP-BY-STEP 5.1 Continued

2. Zoom in on the white card so that all you see is white. (You can use a piece of paper that does have some text on it, but make sure that when you white balance you see nothing but white.

3. Push the button to activate the white balance.

That is all you need to do. It is simple and it certainly makes the video look a lot better.

About Camera Supports

A camera support is simply a way to keep the camera steady without having to hold it with your hands. Most people think of a tripod when they hear camera support, but there are many different types of supports.

A tripod is simply three legs with a head. The head connects the camera to the tripod and allow you to control the camera's position. A fluid head is a tripod head that is filled with fluid that helps make camera moves smooth. You can use a non-fluid head to make your camera moves, but the camera probably will jerk a little as you move. The more fluid there is in the head, the more resistance to movement. If there is too much fluid it is difficult to tilt or pan, if there is too little fluid the head moves too easily. You can adjust the amount of fluid, usually by turning a dial. The fluid, by the way, stays in the head—you can't add fluid.

> **Note**
>
> Don't leave the camera on the tripod when you're not around. Not long ago a student put a $50,000 camera on a tripod and turned around to grab something. By the time he turned around the camera had taken a nose dive. The $8,000 lens and $50,000 camera were both ruined. He wasn't even five feet from the camera. Sometimes it's user error and other times its tripod failure, but make sure you don't drop your camera.

Studio cameras often are mounted on pedestals. Pedestals are similar to tripods, but they don't have three legs. Pedestals add the ability to move the camera and its support on wheels. A camera operator often will be told to pedestal in, or ped in, meaning push the camera support toward or away from the subject.

A second type of camera support is a dolly. A dolly is basically a platform with wheels. Expensive, Hollywood-type dollies have a pneumatic camera support that moves up and down that is built onto the dolly. There are also smaller, less complicated dollies that are more like flatbed wagons that you attach a tripod to. A dolly often is run along tracks to control the position. (The dolly grip is the grip that pushes and pulls the dolly around.)

A third type of camera supports are cranes or job arms. Both of these types of supports allow the camera to get high above the subject and swoop down and move around in different ways. When you see camera moves that seem to sore or fly it probably was done with a jib or crane.

A crane is a large camera support with a large arm that raises a camera, camera operator, assistant camera operator, and maybe a couple other people high above the subject. In short, a crane is a big piece of equipment.

A jib arm is basically a small crane that holds nothing but the camera. Large jib arms require somebody to move and control the jib while the camera operator controls the camera. Small, DV camera-specific jibs allow the camera operator to control both the jib and the camera. A jib arm also can be attached to a dolly to offer even more mobility.

A third type of camera supports are steady cams. Steady cam supports allow the camera operator to walk with the camera as if they were holding the camera by hand. A steady cam removes the shaky, uncontrolled camera motion and gives a steady (therefore the name steady cam) image. Professional steady cam supports are usually pretty expensive, but I have seen plans for homemade steady cams for DV cameras on the internet, and they seem to work pretty well too.

Make sure you use a camera support when you are shooting. I don't like getting seasick watching my handheld camera work and I don't know of many people that like that look anyway. I don't know many people who can afford a dolly or jib arm, but be creative. Use a wagon or wheel chair as a dolly. I once used two office chairs as a dolly and it worked out surprisingly well. Be creative.

There are times, however, when it will be impossible to use a camera support. Let me offer some suggestions for those times.

First, use both hands on the camera. This keeps the camera steadier than using just one hand and you can hold it longer.

Second, lean your body against a wall, or a post, or something else solid. This will help you keep your balance and stop any sway in your body. You also can rest your elbows on a table, desk, or some other solid object.

Third, keep the movement to a minimum. The more you move, the more opportunity there is for the images in the film to look jiggly.

Many cameras have some kind of a image steadying feature. The camera basically crops the image and eliminates the edges. This can cause a problem, however, if the subject is on the edge of the image.

STEP-BY-STEP 5.2

Most professional camera tripods with fluid heads have adjustments and counter balances that allow the operator to perform smooth camera moves and keep the camera still when the camera is tilted up or down. Setting the camera up correctly allows the user to get the most out of the equipment. All heads are different, however, so read the instructions or user's guide to make sure you set the head up correctly. This Step-by-Step covers the basic steps for setting up a tripod head.

1. Center the head on the tripod. Most cameras have a bubble that shows when the head is centered— you simply roll the head until the bubble is in the center.

2. Remove all fluid from the tripod head and release all locks before placing the camera on the tripod. Many tripod heads have dials that show the amount of fluid being used. The tripods I use, for example, have dials numbered from 0 to 10 that show how much fluid is being used for the tilt and pan. The head should move loosely.

3. Place the mounting plate on the tripod and then the camera on the plate. Some mounting plates require you to mount the camera on the plate before you connect it to the tripod.

STEP-BY-STEP 5.2 Continued

4. Balance the camera on the tripod. Professional tripods allow you to slide the camera and camera plate back and forth. Slide the plate back and forth until it is balanced on the top of the head. Make sure the camera has all the weight positioned where it needs to be. For example, some lenses are heavier than others, and batteries are heavier than plugs. Also, stay close to the camera so if it starts to fall you can catch it before it does. Never leave the camera alone.

5. Once the camera is mounted, it is time to adjust the counterbalance if the tripod is equipped with a counterbalance mechanism. The counterbalance keeps the camera still, whether it is tilted up or down.

6. Add the fluid to the tilt and pan. Add enough fluid so that the camera has enough tension for the tilt or pan to move smoothly and under con¡¡trol.

How to Compose Your Shots

A good camera operator knows that the camera decides what the audience sees. A better camera operator knows that the camera can help determine how the audience feels about what it sees. Composition is the way things are arranged in the frame and determines not only what the audience sees, but the mood of the shot. Good composition guides the viewer's eye to the subject of the shot. The principles for good composition in video are the same as they are for still photography and painting. The guidelines we will look at here are:

- The rule of thirds
- Color
- Mass
- Lines
- Framing
- Simplify
- Depth of field

Rule of Thirds

The first basic principle for good composition is called the rule of thirds. Divide the frame into nine equal size cubes with an imaginary tic-tac-toe board. The best locations for good composition are along the top line. The prime locations are where the horizontal and vertical lines intersect (see Figure 5-8).

FIGURE 5-8
The Rule of Thirds

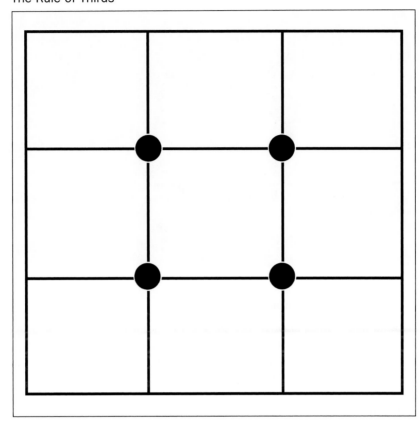

Look at Figures 5-9 and 5-10. The people in Figure 5-9 don't appear to be moving – it looks like they are just kind of sitting there. Figure 5-10, however, makes it look like they just came into the frame and have somewhere to go. Placing the subject in one of the prime locations creates tension and movement even in a still image.

FIGURE 5-9
Subject looks motionless

FIGURE 5-10
Using rule of thirds to create motion

Which of these two pictures is more interesting?

Figure 5-11 looks like a mug shot in the post office, or that the subject's head is stuck in a box. The subject has no depth or drama. Figure 5-12 has depth. The subject appears to be looking at something out of the frame, such as another person. The framing creates a little tension by making us wonder who or what he is looking at.

FIGURE 5-11
Subject's head has no depth

FIGURE 5-12
Subject's head has depth

Does Figure 5-12 use the rule of thirds? The answer is yes. The subject's eyes are in the upper third of the frame. News anchors often are shot in the center of the screen to pull all of the attention to them, but there isn't much visually interesting information in the frame anyway. Center framing works for news anchors because the anchor wants to appear to be talking directly to whomever is watching.

Using the rule of thirds, however, is not fool-proof. Make sure thought goes into every shot.

The subject in Figure 5-13 looks like he is ready to walk into a wall or the picture is going to fall to the left. The photographer followed the rule of thirds, so what is wrong? When framing a shot make sure that you leave plenty of head room and nose room, but not too much of either.

FIGURE 5-13
Rule of thirds without nose room

Nose room is the amount of space in front of the subject's face. Head room is the amount of space above the top of the subject's head.

Figure 5-14 shows too much nose room. The subject isn't really even in the frame and the picture is unbalanced.

FIGURE 5-14
Too much nose room

The subject in Figure 5-15 has too much head room, and most of the frame is empty.

FIGURE 5-15
Too much head room

The subject in Figure 5-16 does not have enough head room.

FIGURE 5-16
Not enough head room

When working with motion, such as a car driving down the road, nose room is referred to as lead room. Figure 5-17 doesn't have enough lead room. The audience is waiting for the people to hit the edge of the frame. Figure 5-18 has enough lead room. The audience feels like the people still have somewhere to go.

FIGURE 5-17
No lead room

FIGURE 5-18
Good lead room

Color

Another way to guide the viewer's eye is through color. Bright objects grab the attention of the human eye. Look at Figure 5-19. The eyes are pulled toward the brighter area.

FIGURE 5-19
Eyes are pulled to the brighter area

Using color also helps balance an image. Figure 5-20 is unbalanced – all of the image is on a single side. Using a darker area in the frame, like Figure 5-21 balances the image so that the view does not feel like the person is going to fall off the side of the frame.

FIGURE 5-20
Image is unbalanced

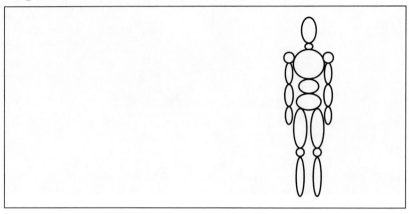

FIGURE 5-21
A darker color balances the image

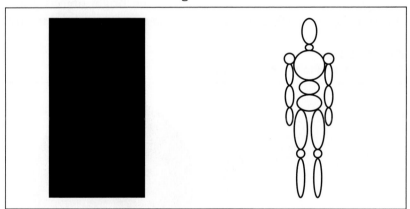

Mass

The human eye notices big things first. Making the subject the largest thing in the frame draws the eye to it. There are two circles in figure 5-22, but we look at the biggest circle first.

FIGURE 5-22
The circles are the same color, but the eyes are drawn to the larger circle

Lines

The human eye likes to follow lines. We see an arrow pointing at something and we look to see what the line is pointing at. Use lines to lead the viewer's eyes to the main subject. The lines in Figure 5-23 lead the viewer's eyes away from the subject. The lines in Figure 5-24, however, point right to the subject. A viewer's eyes may wander, but they follow the lines back to the subject quickly.

FIGURE 5-23
Lines draw the eyes away from the figure

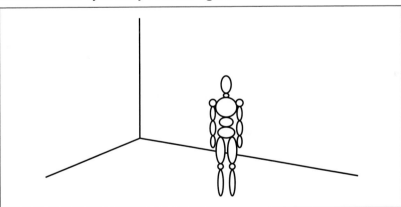

FIGURE 5-24
Lines help draw the eyes to the figure

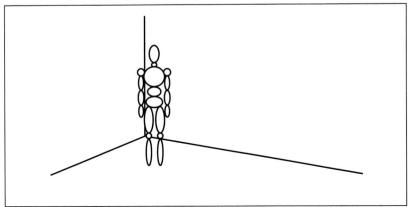

Framing

Another way to direct the viewer's eyes to the subject is to create a frame within the frame. Things like a doorway or a window can be used to direct the viewer's eyes to the subject. Tree branches, walls, plants, etc. can be used to direct the viewer's eye to the subject (Figure 5-25).

FIGURE 5-25
Example of Framing

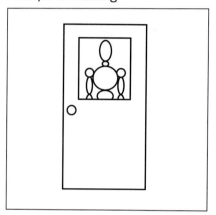

Framing can also be faked. Have a friend or crew member stand just out of frame with a tree branch and hold it up so it that it creates a frame for the image.

Simplify

It is more difficult for a viewer to pick out the subject of an image if there are a lot of other things in the frame to look at. Look at Figure 5-26. The subject gets lost in all of the bright, extra stuff in the background and the viewer's eyes can't stay with the subject. Eliminate the stuff in the background and simplify what you are shooting against. Don't opt for a white wall, however, unless you have a reason for using a white wall. Make the background as interesting as possible, but make sure it doesn't confuse the viewer about what he or she is supposed to be looking at.

FIGURE 5-26
Figure gets lost in the background

Another thing to consider when looking at backgrounds is possible lines, like posts and signs. For example, make sure you don't place the talent right in front of a post or a telephone pole so it looks like he or she has something growing out of their head.

Depth of Field

We talked about depth of field earlier in the lesson, but only about how to control depth of field. Depth of field is another way of directing the viewer's attention. In almost every movie, for example, there is a shot where a person in focus is talking to somebody that is out of focus. When the person in focus stops talking, the focus changes to the person he or she was talking to. This leads the viewer's eye from one subject to another.

Using depth of field also can help you get around non-simplified backgrounds. By narrowing the depth of field you can shoot against almost any background you may come across. A dimly lit fuzzy background is not a bad background at all.

I watched a couple of big budget movies on DVD while I was working on this lesson. I was amazed at how often the rule of thirds was applied, how depth of field was used, and how the angle of the camera affected the mood and feeling of the scene. Pay attention to every detail.

About Field of View

Field of view refers to how much of the subject is seen in the image. A close up (CU) is the subject framed from the chest up as seen in Figure 5-27. An extreme close-up (XCU) is even closer than that. Notice the use of the rule of thirds and that the subject has plenty of nose room.

FIGURE 5-27
Close up

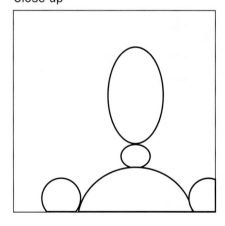

The subject in Figure 5-28 appears to be a disembodied head floating in a box. The audience gets the idea that there is more to the subject in Figure 5-29 because the lines of the head lead the viewer out of the frame.

FIGURE 5-28
This extreme close up cuts off the top of the head

FIGURE 5-29
This extreme close up leads the viewer to believe there is more just out side the frame

A medium shot (MS) is where the subject is seen from the waist up, as seen in Figure 5-30.

A long shot (LS) is a full body shot (Figure 5-31), while an extreme long shot (XLS) is even further away from the subject and shows much more of the surroundings (See Figure 5-32). You learn more about when and why to use all of these different shots in Lesson 7.

FIGURE 5-30
Medium shot

FIGURE 5-31
Long shot

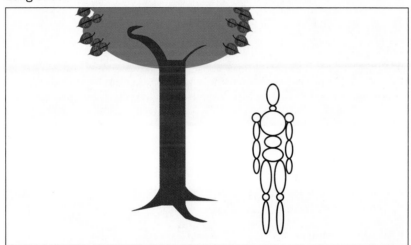

FIGURE 5-32
Extreme long shot

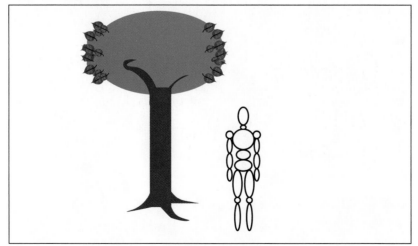

Angle

We live in a three dimensional world, but video images are flat and two-dimensional. Changing the camera angle helps create depth and interest. Figure 5-33 is flat, but changing the camera angle creates depth in Figure 5-34.

FIGURE 5-33
Mugshot has no depth

FIGURE 5-34
Angle creates depth

The camera angle also can add to the mood and feeling of the shot. Generally an eye level shot creates a sense of equality between the audience and the subject, while a high camera angle, which is accomplished by placing the camera above the subject and shooting down at the subject, creates the sense that the subject is small and powerless. A low camera angle, which is accomplished by placing the camera below the subject and shooting up, creates the sense that the subject is large and powerful.

About Camera Moves

In most situations motion should be created by the subject and the camera should stay in one location. Let the actors do the work. There are situations, however, when moving the camera is appropriate. Camera moves are used to reveal more information in the scene. A **tilt** is a camera move from up to down or down to up (Figure 5-35) to show how big the mountain is compared to the character that is getting ready to climb it.

A **pan** is when the camera turns from right to left or left to right. The camera support doesn't move (see Figure 5-36).

FIGURE 5-35
Tilting moves the camera up and down

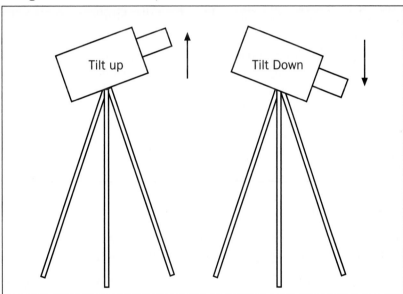

FIGURE 5-36
Panning moves the camera right and left

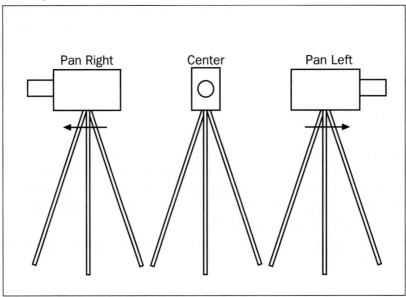

A dolly move is when the camera is moved by the dolly. Studio cameras have a similar move that is called a pedestal move.

Note: These principles are good starting points, but they are guidelines, not hard and fast rules. The composition for each image depends on what looks good, not a bunch of guidelines. It is okay to ignore the guidelines, but make sure you understand them first and know why you are breaking them. Be creative, but start out by being creative within the guidelines we just covered.

SUMMARY

In this lesson you learned:

- A video camera has four basic parts that work together to acquire video. The four parts are the lens, the CCD (or imaging device), storage, and the viewfinder.

- The lens gathers the light waves and focuses the light waves onto the imaging device, which is a CCD in a video camera.

- Different types of lenses perform different functions, with the function determined by the focal length. A 50mm lens is a normal lens, which mimics what we would see with the normal eye. A telephoto lens has a larger focal length and brings objects closer to the camera. A wide angle lens has a shorter focal length and a larger field of vision than a normal lens. A zoom lens has a variable focal length and can serve as all three of the other types of lenses.

- Composition refers to how things are arranged in the frame. Good composition guides the viewer's eye to the subject. Guidelines for good composition include the rule of thirds, color, mass, lines, framing, and depth of field.

- Field of view refers to the amount of the world that is captured by the camera. The field of view is referred to by extreme close-up, close-up, medium shot, long shot, extreme long shot.

- Camera supports help the camera stay stable. Camera supports include tripods, pedestals, dollies, jib arms, cranes, and steady cams.

- Camera moves include pans, tilts, dolly moves, and jib arm moves. Camera moves are used to reveal more visual information.

VOCABULARY *Review*

Define the following terms:

Aperture	Field of view	Nose room
Camera	Focal length	Optical zoom
Charge-coupled device (CCD)	Focal point	Overexposure
Composition	Framing	Pan
Convex	Head room	The rule of thirds
Depth of field	Imaging device	Tilt
Digital zoom	Lens	Underexposure
Exposure		

REVIEW *Questions*

MULTIPLE-CHOICE QUESTIONS

1. What does focal length measure?
   - A. the total length of a lens
   - B. the diameter of the aperture when the image is in focus
   - C. the length from the front of the lens to the back of the camera
   - D. the distance between the lens and the imaging device

2. If a lens has a focal length of 100mm and the aperture is set at f/2, what is the diameter of the aperture?
   - A. 12.5mm
   - B. 25mm
   - C. 50mm
   - D. 100mm

3. What does a telephoto lens do?
   - A. brings the subject closer to the camera
   - B. has a wider field of view
   - C. has a variable focal length
   - D. is known as a zoom lens

4. If the lens has a focal length of 50mm and the aperture is set at f/4, what is the diameter of the aperture?
   A. 100mm
   B. 15mm
   C. 12.5mm
   D. 25mm

5. What happens when you increase depth of field?
   A. move the camera closer to the subject
   B. move the camera further away from the subject
   C. increase the aperture
   D. open up the aperture

6. What happens when you decrease depth of field?
   A. move the camera closer to the subject
   B. move the camera further away from the camera
   C. decrease the aperture
   D. close-down the aperture

7. Each CCD in a three CCD camera _____.
   A. is sensitive to the same things
   B. is a different size
   C. is sensitive to a different color of light
   D. captures the red, green and blue light

8. How are the different colors of light separated in a three CCD camera?
   A. by the lens
   B. by a series of prisms
   C. by the CCD
   D. the light colors are not separated

TRUE/FALSE QUESTIONS

Circle T if the statement is true or F if the statement is false.

T   F   1. The imaging device in a video camera is called a charge-coupled device.

T   F   2. A convex lens is wide on the edges and thin in the center.

T   F   3. A single CCD camera produces a higher quality image than a three CCD camera.

T   F   4. A jib arm is basically a small crane.

T   F   5. A dolly is a tripod on wheels.

T   F   6. A pedestal is basically a tripod on wheels.

T   F   7. White balance refers to altering the color temperature of the light source.

T   F   8. A lens over 50mm is considered a wide angle lens.

T   F   9. Composition refers to how the objects in the frame are arranged.

T   F   10. The tripod head connects the camera to the tripod.

SHORT ANSWER QUESTIONS

Write a brief answer to the following questions.

1. Explain how to white balance the camera.

2. Describe how a video camera works. Make sure to discuss the four parts of the camera.

3. Describe the rule of thirds.

4. Describe three ways to help stabilize a hand-held camera.

PROJECTS

PROJECT 5-1

Obtain a camera and figure out how to use the white balance. Practice so you can white balance easily and quickly when you have to. Notice the difference in the image before you white balance and after you white balance. In some instances you may not notice much of a difference, but in most situations you should see a big difference.

PROJECT 5-2

One of the best ways to learn how to compose a shot is to look at photographs and paintings and watch movies, lots of movies. But when you watch movies and look at pictures do it to learn. Watching movies is a great, inexpensive film school. While it is important to study great works of art, it is just as important to study not so great works as well. Watch what you can as often as you can and ask questions. Where is the subject in the image? How is the picture composed? Does the artist use the rule of thirds? If he or she doesn't use the rule of thirds, why? Does the image work? Why or why not? What is the field of view? How does the field of view affect the mood of the scene? You would be surprised to find that the guidelines just covered are used more often than not.

PROJECT 5-3

Grab the camera and go shoot some video. Experiment with the composition guidelines. Change the field of view and throw in some camera moves. The idea of this assignment is to spend quality time with your camera. Make sure to experiment with everything we talked about. Compare what you shoot holding the camera with one hand to what you shoot when holding the camera with two hands. Change the white balance. Get to know the camera by both using it and reading the user manual. There is a lot of general information about video in a user manual. Then study what you shot. What can you improve? What do you do well? What do you like? What don't you like?

 ## WEB PROJECT

Search the Web for paintings from Rembrandt and Michelango and see how they use the composition principles that we have discussed in this lesson: the rule of thirds, exposure, depth of field, white balance, angle, color, mass, lines, framing, and field of view. Write your observations in a word processing document.

 ## TEAMWORK PROJECT

Divide into groups of two or three people. Use a digital camera and take some pictures that illustrate the principles discussed in this lesson: the rule of thirds, exposure, depth of field, white balance, angle, color, mass, lines, framing, and field of view. Show your pictures to other members of the class and compare your pictures with the other groups. Analyze which pictures are good or poor and why.

CRITICAL*Thinking*

ACTIVITY 5-1

Look at some art work, pictures, and videos and see how you can improve the composition and camera work. Is there too much camera movement? Is there too little camera movement? Is it overexposed or underexposed? Does the composition help the emotion of the story? Does it improve the story? Does it hinder the story?

LIGHTING

VOCABULARY

Arc light
Backlight
Barn doors
Butterfly
C47
Candelas
Color temperature
C-stand
Cucaloris or cookie
Diffusion
Dimmer
Direction
Fill light
Flags
Foot-candle
Fresnel
Gel
Gobo
Halogen
Hard light
Intensity
Key light
Kelvins
Lux
Neutral density (or ND) gel or filter
Overhead
Reflector
Scrim
Set light
Silk
Soft light
Tungsten
Wattage or Watt

Introduction

In the last lesson we learned how to use a camera and how to compose shots. However, there is another element to shooting good video: lighting. Good lighting is basic, but it can take a lifetime to master the basics.

Basic lighting ensures there is enough light on the subject for the camera to be able to expose the image. Better lighting creates shadows and depth. Great lighting does all of that and creates mood, develops emotion, and expresses the theme of the story as well.

In this lesson we talk about what each light is supposed to accomplish, and the tools designed to help you manipulate the light to accomplish what you want.

Identify and Define the Attributes of Light

It is important to discuss the attributes of light before talking about how and where to set up the lights. This lesson focuses on four attributes of:

■ Color temperature

■ Hard versus soft light

■ Intensity

■ Direction

Color Temperature

There are two types of light sources: natural and artificial. The two natural sources of light are the sun and fire. Fire could be considered as both an artificial and a natural light source. Artificial light sources generally rely on electricity to create light.

The difficulty for video work is that the light from these different sources comes in different colors. Light from sunlight produces a blue tint, while light from artificial sources produces orange, yellow, or even green. If the camera is not set up to shoot in the correct color of light, the image looks a little (sometimes a lot) blue, red, green, etc.

The color of the light depends on its **color temperature**. Color temperature is measured in **kelvins**. A candle, for example, burns at around 1200 kelvins and produces a red or orange glow. A professional cinema light is approximately 3200 kelvins and produces a more orange glow than a candle. Sunlight is measured at 5600 kelvins and produces a bluish glow, while sunlight on an overcast day is measured at 7000 kelvins and produces a deeper blue light. Generally, blue is considered a cool color and red is a warm color, but according to the kelvin scale, red is a cool color while blue is a warm color (see Figure 6-1).

FIGURE 6-1
Color temperature scale

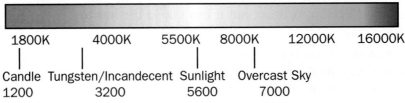

| 1800K | 4000K | 5500K | 8000K | 12000K | 16000K |

Candle   Tungsten/Incandecent   Sunlight   Overcast Sky
1200   3200   5600   7000

In the last lesson we talked about white balancing the camera so the camera knows what white looks like in whatever light you are using. The camera uses different filters to adjust the camera to the light. Shooting using different colored light sources, however, confuses the camera. Shooting with both artificial light and sunlight is generally not a good idea.

When shooting in a room with a window, for example, you should use either the sunlight or cinema lights, but do not mix them unless you have a way to adjust the color temperature of the cinema lights. It is best to

Note

Kelvins are not the same as degrees Fahrenheit or Celsius. If it were, you would get a pretty good burn by just lighting a match. Kelvins compare the light source to what scientists refer to as a "black body." Black body just means something black when it is cold, like a lump of coal. When that black body is heated, it changes color as the temperature increases.

close the blinds or curtains when you choose to use lights, or keep the lights in the case and use sunlight. You can, however, change a cinema light's color. If there is not enough sunlight, for example, you can use a cinema light to add to the sunlight by placing a blue **gel** in front of the light. The blue gel changes the color of the light to something closer to sunlight. Some lights are created to match the color of sunlight. You also can use color compensation filters (often they are built into the camera) to adjust the color temperature.

> **Note** ✓
>
> Filters are a lot of fun to work with. You can "warm up" an image by using a gold filter on an interview, or remove glare from water or glass by using a polarizing filter. We won't discuss filters right now, but if you are serious about videography or cinematography you will want to learn about filters.

Hard Versus Soft Light

Walk around outside on a clear, sunny day and look at all the shadows. The shadows are hard and well defined. Hard shadows are not particularly appealing, but they are dramatic and powerful. Notice, as well, that each object has only one shadow, and all of those shadows fall in the same direction. The sun has a monopoly on shadow creation outside and the position of the shadow is determined by the sun's position in the sky. Light that produces hard shadows is known as **hard light**. The sun is a small light source (it is far away from the earth which makes it a small light source) that comes from one point to produce hard shadows (see Figure 6-2).

FIGURE 6-2
Example of a hard shadow

Soft light sources, however, create soft, less defined shadows. Soft shadows are more appealing because they soften the edges of an object. Soft light sources are generally larger light sources, where the light is spread out and hits the subject from more than a single point. Watch the news when you get chance and pay attention to the shadows. Most lighting for documentaries and studio news broadcasts use soft light, as well as those romantic scenes in movies (see Figure 6-3).

FIGURE 6-3
Example of a soft shadow

Some lights are made specifically to produce soft light, while other lights produce hard light. The hard lights, however, are not doomed to produce only hard light. Hard light can be softened. Diffusion softens the light and the shadows the light creates. Diffusion spreads the light out so that a single source, like the sun, hits the subject from more than a single point. One method of diffusion is to bounce the light off the wall, ceiling, or light reflectors. Another method to diffuse light is to put something in front of the light, such as silk or an egg crate. We will discuss these diffusion tools a little later on.

The decision to use hard or soft light depends on the mood or emotion of the scene. A romantic scene, for example, might call for soft light and soft shadows. A bank robbery, on the other hand, is more dramatic and may call for hard light and hard shadows.

Intensity

As discussed in Lesson 5, the video camera relies on light to create images. If there is too much light, the image is overexposed. If there is not enough light, the image is underexposed. Getting the right quantity of light is very important. The strength of light is called intensity. The more intense the light, the more light you have. Light intensity is measured in **candelas**, **lux**, and **foot-candles**. These three measurements determine how much light falls on a specific area from a specific distance.

The intensity of a light can be reduced using a few methods. The first method is to simply add distance between the light source and the subject. The closer the light is to the subject, the more intense the light is.

A second method for reducing the intensity is to use a **dimmer**. A dimmer reduces the flow of electricity to the bulb. The problem with using a dimmer is that the dimmer changes the color temperature.

The third method is to us a **scrim**, which we will discuss more in-depth later on. A scrim is metal screen placed in front of the light. A scrim reduces the intensity without changing the hardness of the light. Scrims are good for when you do not have room to move the lights or the subjects around. Just remember to use scrims designed specifically for lighting; otherwise, you might start a fire (see Figure 6-4).

> **Note**
>
> You will want to know more about candelas, lux, and foot-candles if you choose to become more serious about video lighting, but for now just know that you will come across these measurements as you go along.

FIGURE 6-4
Scrim being placed on a light

Photo courtesy of Lowel-Light Mfg., Inc

The fourth method is to use an **ND gel**. ND stands for neutral density. A gel is made specifically to work with the heat produced from lights without burning or melting. Some gels can change the color of the light, such as blue gels that help match the color of lights to sunlight. An ND gel, however, reduces the intensity of a light without changing the color. ND gels are identified with numbers—a .9 ND gel reduces the light more than a .3 ND gel, while a .6 ND gel is somewhere in between (see Figure 6-5).

FIGURE 6-5
Variety of gels

Darker gray gel
is an ND filter
and reduces
the light
intensity

Lighter gray
gel is a frost,
which diffuses
and softens
the light

Blue gels
alter color
temperature

Photo courtesy of Lowel-Light Mfg. Inc.

The last way to alter the intensity is to use a bulb with a different wattage. A watt measures the number of electrons that are moving down a wire to a light fixture or bulb. More watts generally equal greater intensity. While a lower watt bulb will produce a lower intensity, it is not always as easy as it would seem. A 100-watt bulb does not produce twice as much light as a 50-watt bulb. Most lighting equipment manufacturers provide information about how much light each bulb supplies.

Direction

Each light produces a shadow. The position of the shadow depends on the direction from which the light is coming. Controlling the shadow is pretty simple when you recognize the direction of the light. A light above the subject creates an eye shadow on someone with deep-set eyes, while a light positioned on the right side of a subject will create a shadow on the left side of the subject (see Figure 6-6).

FIGURE 6-6
Light direction

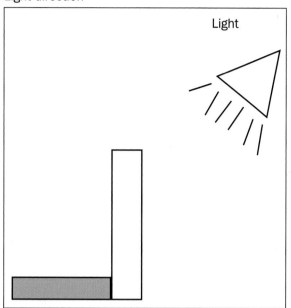

S TEP-BY-STEP 6.1

In this Step-by-Step activity you pay attention to the light and shadows around you.

1. Open a word processing program and create a new document.

2. Answer the following questions:
   a. Where is the light coming from?
   b. Is it the window? Is it the light overhead?
   c. Where are the shadows?
   d. Move the light around if you can and observe the shadows. Have the subject move around in the same light.
   e. How does that affect the shadows?
   f. Are the shadows hard or soft?
   g. What kinds of lights are creating what kinds of shadows?

3. Save the document as SBS6-1.doc and close the application.

Hot Tip

Often you find you have shadows and you can't figure out from where the light is coming. Turn the lights off one at a time if you can, or wave your hand between the subject and where you think the light is coming from. If your hand creates a shadow where you think it should, you can identify the source.

Identify the Types of Artificial Light Sources

Most lights found on film sets and in television studios are expensive and require a lot of special equipment to use and maintain. Fortunately more and more companies recognize the impact of DV and are producing high quality lighting equipment at more reasonable prices.

The most common lights are Tungsten lights. Tungsten, which is a metallic element, is used to create filaments that generate light through the production of heat. Tungsten filaments create a consistent color temperature of 3200 kelvins, which produces a yellowish color. However, the tungsten filament is destroyed over time because of the heat. Halogen inside the bulb helps slow that destruction. The case surrounding the Tungsten and Halogen is made of quartz, so you also hear about Tungsten-Halogen and Tungsten-Quartz bulbs. It is important to remember not to touch a bulb with your bare hand because the oils in human skin shorten the life of the bulb.

Tungsten lights also are known as incandescent lights and come in a couple of different types. The first type is enclosed spotlights, the most common being the Fresnel (the "s" is silent, so it is pronounced Fren-el). The name Fresnel comes from the lens that directs the light and makes it either a spotlight or floodlight. Fresnel lights have been around for more than 600 years since the Fresnel created lights for lighthouses. However, in 1935, an Academy Award was given for the Fresnel Solar Spots and are probably the most common studio lights.

The second type of Tungsten light is an open-face light, which is simply a bulb with some type of reflective material behind it. Some open-face lights are hard light sources, while others are scoop lights that create soft light. Many of the smaller, more portable lights used in DV are open-face lights that come with a variety of light controls.

Fluorescent lights, like the ones in most schools and offices, are horrible for film and video because of inconsistent color temperature. Fluorescent lights manufactured specifically for film and video lighting have a very consistent color temperature. Many fluorescent light kits also come with daylight tubes with the same color temperature as daylight. This allows you to use mixed light sources.

The last type of light are arc lights. Arc lights are generally bigger, harder to move around, and require specialized generators or power supplies. These types of lights are good for big, Hollywood type shoots, but are usually too big for someone shooting on DV.

Light Control

I often tell clients that film and video people want to control sound and tell people what to do and say. We even think we can control the sun. We cannot really control the sun, but with the use of a few tools we can soften the light of the sun and keep the light from shining where we do not want it. Artificial lights are even easier to control. Placing cloth, wood, metal, or other objects between a light source (even the sun) and a subject gives a videographer or cinematographer control of where light shines and what the light looks like on the subject. In this section we talk about different light controls you can use.

One of the easiest ways to control where a cinema light shines is to use barn doors. Barn doors are hinged shutters that can be moved into place over the light. Barn doors do not decrease the intensity, but they control where the light shines (see Figure 6-7). For example, if we want the light to shine on the right side of a subject's face we can close the barn doors to create a block between the light and the specific side of the subject. Unfortunately, there are no barn doors for the sun.

FIGURE 6-7
Example of barn doors

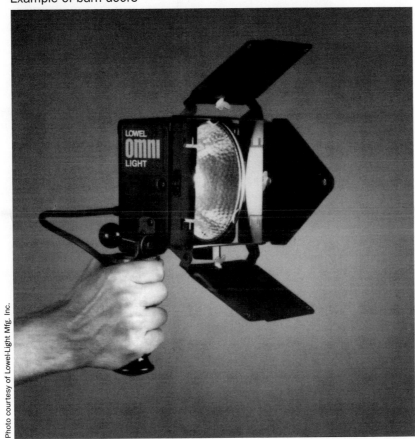

Photo courtesy of Lowel-Light Mfg. Inc.

You can use a flag to control where the sun or other light sources fall. **Flags** are opaque pieces of material that do not allow light to pass through. Some flags are made of cloth that fit into frames of various sizes, but others are made of metal, and you could even use a piece of wood if you wanted. Just remember that cinema lights are hot, so do not put anything too close to them. I have melted or burned all sorts of stuff with cinema lights, so be careful (see Figure 6-8).

FIGURE 6-8
Flag

Photo courtesy of Matthews Studio Equipment, Inc.

We talked about scrims earlier but in the context of placing them onto a light. Another type of scrim is a net-like material put into frames of various sizes so you can use it for any light source, including the sun. The light shines through the empty spaces in the scrim and reduces the intensity. A double scrim reduces the intensity even more. A half scrim is used to knock down only one side of the light, allowing the full intensity to hit half of the subject. Imagine two objects side by side in a shot. The first object is overexposed, but the second object is just right. Use a half scrim to knock down the amount of light on the overexposed object while allowing the original amount of light to fall on the second object. You also can use a half scrim to keep a moving subject from becoming overexposed when they move closer to the light source.

Silk refers to a piece of artificial silk that allows some, but not all, of the light onto an object. A silk can be used between the light source and the subject to diffuse and soften light (see Figure 6-9).

FIGURE 6-9
Silk

Photo courtesy of Matthews Studio Equipment, Inc.

Cucaloris, also known as a cookie, is simply a piece of opaque material (often wood) with shapes or patterns cut out (see Figure 6-10). Part of the light is blocked, while the shapes or patterns allow light to pass through. You also can use plants, branches, or what ever else you want to create a cookie. Cookies generally cast soft shadows. A gobo is similar to a cookie in that it shapes light. While a cookie creates non-specific shapes, a gobo creates a specific shape or pattern like a house, a witch, or prison bars. A gobo usually creates harsh, well-defined shadows, while a cookie creates more diffuse patterns.

FIGURE 6-10
Cucaloris

Photo courtesy of Matthews Studio Equipment, Inc.

Reflectors are used to reflect light back onto a subject. This allows you to get double duty out of a single light source. One of my favorite methods of lighting is to use a single light source and reflect the light to where the shadows are darker than I want them to be. You can buy photographic or cinema reflectors, or simply use a piece of white foam core, which is basically Styrofoam sandwiched between two pieces of white construction paper.

Two other light controls are overheads and butterflies. These light controls are good for any situation, but come in really handy when you are working in daylight. Butterflies and overheads are large frames to which you can attach silks or scrims and use to control light coming down onto the subject. Butterflies generally use a single stand, while overheads require two big stands. Butterflies and overheads perform the same duties and generally control light over larger areas or surfaces (see Figure 6-11).

FIGURE 6-11
Butterflies and overheads

Photo courtesy of Matthews Studio Equipment, Inc.

Many of these light controls require stands or clips to put them into position. The most common stand is a **C-stand**. This stand comes with an extension arm and a multi-position head that allows the grip, or the person who moves things around and puts them where they need to be, to position the light control where it is needed (see Figure 6-12). Most light manufacturers create a variety of stands to use with their lights, and you can be creative with how you use them to hold any of the light controls we have covered.

FIGURE 6-12
C–stand

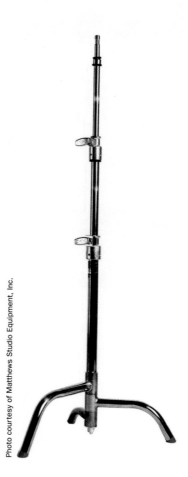

Photo courtesy of Matthews Studio Equipment, Inc.

Most manufacturers make some kind of clip to connect gels and other light control equipment to lights, frames, and barn doors. The most common and least expensive are C47s. C47 is just a fancy name for a clothespin. Make sure, however, to use only wood C47s because plastic ones will melt.

Three-Point Lighting

It is possible to light a shot using just one or two lights and make it look good, but the most common lighting setup is known as three-point lighting. Three-point lighting uses three main lights: the key light, the fill light, and the backlight. Sometimes a fourth light is used as well, called the set light.

Three-point lighting is used in many situations, including interviews and news. Good three-point lighting is pleasing to look at and creates flattering shadows when done properly. Three-point lighting is not the most basic, but it is a starting point for more advanced lighting. Let's look at each light individually.

The Key Light

The first light you need to set is the key light, or simply the key. The key is the main source of light and creates the main shadows on the subject. You can place the key to the side of the subject, below the subject, above the subject, or wherever you want. When following three-point lighting, however, the rule of thumb is to place the key light at a 45 degree angle to the side of the subject and a 45 degree angle down from the subject. (See Figure 6-13). You may want to adjust the position of the key light, depending on the mood you are trying to create, but starting at 45 degrees gives you a good starting point.

Another rule of thumb is to place the key on the side opposite the direction the subject is looking. If the subject is looking toward the right, the light from the key should be on the left side of his or her face (see Figure 6-13).

FIGURE 6-13a
Positioning the key light

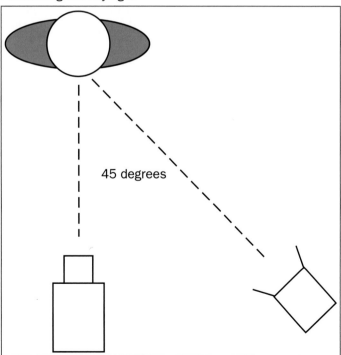

45 degrees

FIGURE 6-13b
Positioning the key light

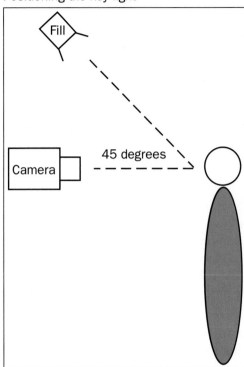

Once you have set the key take a look at the shadows created. The shadows should help you decide where the key needs to be. If the subject has a long nose, for example, the shadow from the key may make the nose look even longer. Moving the light closer to the camera would reduce the length of the shadow created by the nose.

The Fill Light

The second light you need to position is the **fill light**. The fill light fills in the shadows created by the key while not creating any of its own shadows. Filling the shadows created by the key does not mean eliminating the shadow, but it should lighten or soften the shadow depending on the effect and mood of the scene. The rule of thumb for placing the fill light is to position it on the side opposite the key light and closer to the camera. It also should start out about level with the camera lens (see Figure 6-14).

FIGURE 6-14
Positioning the fill light

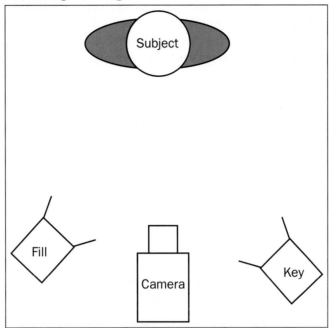

One of my favorite ways to create a fill light is to use a reflector. The reflector uses the light from the key to fill in the shadows created by the key. The reflector may need to be placed close to the subject, but good framing (how things are positioned in the shot) makes sure the reflector does not appear in the shot.

The difference between the intensities of the key light and fill light is called the lighting ratio. If the key and fill lights have the same intensity they have a ratio of 1:1. If the key light is twice as powerful as the fill light the ratio would be 2:1, with the key light being the first number (2) in the ratio. The higher the ratio, such as 4:1, the darker the shadows are when compared to the light areas.

High-key lighting has a lower ratio, such as 2:1, and uses bright, diffused light to create soft shadows. Most television news and documentary interviews are shot in high-key lighting. High-key lighting is appealing, but not dramatic. A higher ratio, 4:1 for example, creates low-key lighting with more dramatic contrast between the light and dark areas. The lighting ratio depends on the emotion needed in the scene.

The Backlight

The third light is the backlight. The backlight's job is to create a halo of light that outlines the subject's edges. The backlight creates separation between the subject and the background. The backlight should be placed behind the subject and pointed down onto the subject. The backlight may need to be brighter (have greater intensity) than even the key light in order to do its job (see Figure 6-15).

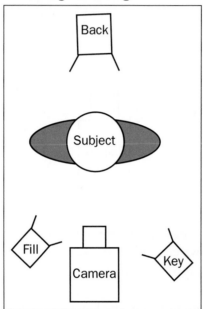

FIGURE 6-15
Positioning the backlight

One of the difficulties with the backlight is that if the backlight isn't placed carefully it may create new shadows on the face or body of the subject. Lighting controls, like barn doors and flags, can help you control where the light will shine.

The Set Light

The last light is the set light. The set light is directed at the set to create separation between the subject and the background. The set light should not be so bright that it pulls the attention away from the subject, but it needs to be bright enough that there is separation between the subject and the set. A gobo or cookie can be used to create texture, depth, and interest in the background (see Figure 6-16).

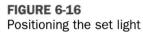

FIGURE 6-16
Positioning the set light

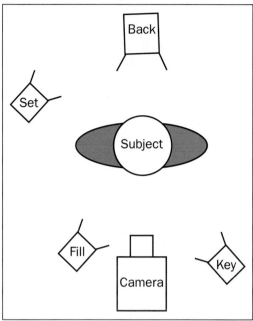

Three-point lighting also can be used to light for more than one subject. The principles are the same, but it may require more lights. A key light can be set for each subject, or a single key light can be used for all of the subjects, depending on where they are in relation to each other. The fill light for one subject may add a strange shadow on one of the other subjects. Take your time and use flags and barn doors to make sure you get it right. Remember, do not be afraid to turn off all of the lights and look at each light alone to see what is going on.

Sometimes, when I get stuck in a lighting situation I go back to the idea that I am working with shadows instead of lights. When I look at the shadows, I concentrate on putting the lights in the best location to make the shadows look the way they need to look.

A few years ago I had a chance to watch an accomplished cinematographer friend at work. I was excited to see what his lighting secrets were. I watched him set up, check the camera, and adjust the lights. He moved solids and silks around, blocked light, checked it all again, and then finished up. When he was finished, the shot was beautifully lit and much better than anything I could do, but his secret for great lighting was three-point lighting.

Part of the reason my friend's camera work looked so nice was because he worked with people that set lights and knew their jobs as well as he knew his. Another reason was because of all of the nice equipment he had to work with. But the biggest reason his work was so impressive was because he had studied, worked, practiced with lighting and knew exactly what he was doing. He mastered the basic three-point lighting system.

Always remember that there are no hard, fast rules with lighting. Learn three-point lighting, then experiment. Try different things. Lighting is an art and a science, so be an artist with an understanding of the science of how to control light.

STEP-BY-STEP 6.2

In this Step-by-Step activity, you practice the concepts of light control and three-point lighting.

1. Light a subject using three-point lighting. If you do not have access to a light kit use whatever lights you have available, such as candles, flashlights, lamps, etc.

2. Be creative and use the tools you have. Follow the rules of thumb for light placement and then move the lights around.

3. Use reflectors.

4. Try to create different lighting ratios.

5. What effects can you create? How does one light source affect the lighting compared to another?

Lighting Outdoors

Outdoor lighting can be tricky, but the idea of the three-point lighting is the same. It is easy to control the light from a light kit, but the light from the sun is much harder to control.

The easiest way to light outdoors is to use the sun. The problem is that you cannot control where the sun is and you cannot put an ND gel on it to reduce the intensity. However, you can control where the subject is in relation to the sun and you also can use diffusion to keep the sunlight from creating shadows that are too harsh. Even a small silk or scrim on a c-stand makes a big difference.

The best times to shoot outdoors are when the sun is first coming up in the morning or when it is going down at night. The sunlight at these times is usually pretty soft and creates wonderful shadows. The sunlight in the middle of the day is very hard and creates hard shadows, and there is not a whole lot you can do about it.

If you cannot shoot at sunrise or sunset it is possible to shoot outside, but you either want to shoot in the shade or have a few pieces of equipment, like butterflies and overheads. One trick that I have found is to use the sun as a backlight with a silk and then bounce the light back into the subject with a reflector. With a little skill and practice you can make it work.

STEP-BY-STEP 6.3

In this Step-by-Step activity, you experiment with lighting outdoors.

1. Use reflectors or daylight lamps to light a person outside.

2. Position the subject with the sun in different locations in relation to the subject.

3. How does where the sun is affect the subject? Which shadows are more appealing than others?

Lighting Theory

The other night my daughter came out of her room and the left side of her face was lit brightly. The hall was dark and the only light came from her bedroom behind her. Not only was the light coming from behind her, but it was coming from behind her right side, not her left. How come the left side of her face was lit, not her right side, and how did it get in front of her?

I figured out that the light bounced off the wall at my daughter's left side and back into her face. I put a blanket against the wall to make sure I was right. I should have known that the light was bouncing back into her face because light behaves in a predictable way (for the purposes of lighting for video it does). If light hits a white wall it is going to bounce back and cast the light somewhere else; for example, if two people are facing each other the sun will hit one person on the left side of the face, and will hit the other person on the right side of the face. Remember that the audience expects the light to be in a certain place and behave in a certain way, so you need to make sure you do not try to change the rules.

If the scene is set in front of a fire in the fireplace, for example, the audience expects the light to come from the fireplace. If two people are looking at each other in front of the fire the light should hit one person on the right side of the face and the other person on the left side of the face. A light from directly overhead doesn't make sense in that situation.

Even if the main light comes a single source, that does not mean you would have to use a single light to light the scene. The single light may be to dark to shoot. You then need to use fill light or back-light, but do it in a way that looks natural. Remember, light bounces off of walls and tables, and windows and other objects, so use that idea to make the lighting match the mood or idea from the scene.

The lighting for a scene should be planned out just like anything else. The first thing you will want to do is read the script. Where is the scene set? Where is a logical position for a light to come from? Will the light come through a window? From a lamp? From a fireplace? How many characters are in the scene? Where will each character be located?

Second, visualize. Where do you put the key light for the first character? Will the key for one character work for the second character? Will one light work as the key for the entire scene?

Third, block out the scene. Blocking a scene means you have the actors walk through the scene from beginning to end. The director needs to help the actors know where they should be and when they should be there.

The fourth thing you do is set the lights. The lights may need to be adjusted for the actor's height and face shape. Make sure light stands are out of the shot and that they are weighed down so they don't fall.

The last thing you need to do is white balance the camera. You already learned how to do that from Lesson 5.

STEP-BY-STEP 6.4

In this Step-by-Step activity, you practice using the concept of lighting theory.

1. In the last lesson you were told to look at photographs, paintings, and movies to learn about composition. Now it is time to look at those photographs, paintings, and movies to learn about lighting. Identify where the key light is coming from. Observe your family eating dinner (without being too obvious). Where is the key light in the dinner table scene? Is light being reflected in different locations? What is reflecting the light?

SUMMARY

In this lesson, you learned:

- The two types of light sources are natural (the sun) and artificial (man made).

- Light has four attributes that need to be understood in order to control light: color temperature, hard and soft light, intensity, and direction

- Light is produced in different colors. The color of the light is determined by its temperature, known as color temperature. The color of light is measured in kelvins.

- It is best to avoid shooting in mixed light, which is light with different color temperatures. You can change the color of one of the sources of light by using a gel to change the color at the source, or use a filter on the camera lens.

- Light intensity refers to the strength of the light. Intensity can be reduced in one of five ways: by moving the light away from the subject, using a dimmer, using a scrim, using an ND Gel, or using a bulb of different wattage.

- Light can be either hard or soft. Hard light produces strong, hard shadows like the shadows at noon on a sunny, clear day. Soft light produces soft shadows that generally are more appealing.

- The direction of the light determines where the shadows will fall.

- Three-point lighting uses three (or four) lights to create appealing images: a key, which is the main light source; the fill, which softens or lightens the shadows created by the key; and a backlight to create separation between the subject and the set. The fourth light is the set light that creates texture on the set and creates even more separation between the subject and the set.

VOCABULARY *Review*

Define the following terms:

Arc light	Fill light	Lux
Backlight	Flags	Neutral density (or ND) gel
Barn doors	Foot-candle	or filter
Butterfly	Fresnel	Overhead
C47	Gel	Reflector
Candelas	Gobo	Scrim
Color temperature	Halogen	Set light
C-stand	Hard light	Silk
Cucaloris or cookie	Intensity	Soft light
Diffusion	Kelvins	Tungsten
Dimmer	Key light	Wattage or Watt
Direction		

REVIEW *Questions*

TRUE/FALSE

Circle T if the statement is true or F if the statement is false.

T F 1. All light has the same color.

T F 2. Soft light produces soft shadows.

T F 3. The key light should not create any shadows.

T F 4. C47 refers to a light stand.

T F 5. Sunlight has a blue tint to it.

T F 6. Color temperature is measured in degrees Celsius.

T F 7. Barn doors restrict where light shines.

T F 8. The key light is always in front.

MULTIPLE CHOICE

1. What does intensity refer to?
   A. whether a light is a spot or flood light
   B. the direction of the light
   C. the strength of the light
   D. the size of the light bulb

2. Which of the following options describe light direction?
   A. is not important
   B. controls where the shadows will fall
   C. is determined by strength of the light
   D. is only important for the key light

3. What does a scrim do?
   A. reduces intensity
   B. diffuses light
   C. changes color temperature
   D. creates shadows on a subject

4. What does a neutral density gel do?
   A. reduces intensity
   B. diffuses light
   C. changes color temperature
   D. creates shadows on a subject

5. What is the color temperature of Tungsten lights?
   A. 1200 kelvins
   B. 3200 kelvins
   C. 5600 kelvins or higher
   D. 7000 kelvins or higher

6. What is the color temperature of sunlight?
   A. 1200 degrees
   B. 3200 degrees
   C. 5600 degrees or higher
   D. 7000 degrees or higher

7. What is the rule of thumb for placing the fill light in three-point lighting?
   A. 45 degrees from the center and 45 degrees down
   B. above the subject
   C. behind the subject
   D. closer to the camera than the key light and on the opposite side

8. Which one of the following statements is true about the backlight in three-point lighting?
   A. should be directed at the set
   B. create separation between the subject and the set
   C. should be placed at a forty-five degree down angle
   D. is the main light source

WRITTEN QUESTIONS

Write a brief answer to the following questions:

1. Explain color temperature. Include examples of light sources with the temperatures and colors they produce.

2. Describe three-point lighting and the function of each light.

3. Name and describe three kinds of light controls.

4. Describe three ways to reduce the intensity of a light source. Include advantages or disadvantages to each method.

5. Explain the theory of lighting as discussed in the lesson.

PROJECTS

PROJECT 6-1

Set up lights for a talking head interview using three-point lighting. Place the lights in the positions described in the lesson. Once the lights have been set, re-set the lights, with the lights in different positions, but maintain the same function (key light is still the key, fill light is still the fill, etc.) Which positions look best for the subject? What effect do the different positions have on the subject? Is the lighting more dramatic? Is the lighting less dramatic? Use all of the light controls you have access to, including creating some of your own, such as cutting out a cookie from paper. Compare what you create to still portraits and television interviews. What works? What doesn't?

PROJECT 6-2

Watch a movie, television show, or documentary of your choice. Watch what happens with the lighting from scene to scene or from report to report. Pause as you go (if you can) and try to figure out where the lights are positioned. Where is the key light? Where is the fill light? Where are the lights and how are they used? Is the lighting dramatic? Does it help or hinder the scene or shot? Observe carefully and decide how you would change it and explain why or why not.

 ## WEB PROJECT

Look up different film and television light manufacturers online and download or order a catalog of their equipment. Read through the catalog and figure out what the equipment is and how it is used. Write questions you have about the equipment and then ask those questions to your teacher, or even to the manufacturer. Is some of the equipment the same stuff with different names? Why would you want to use one manufacturer's equipment instead of another manufacturer's?

 ## TEAMWORK PROJECT

Light a scene as a class. If it is for a news show there should be at least two presenters or anchors. If you light for another type of scene, there should be at least two actors. Remember that the characters will move around, so you need to light for their movement. Use the idea that light comes from a main source, but you need to light the scene so that it looks natural and the audience can see what is happening.

CRITICAL *Thinking*

ACTIVITY 6-1

Find images online or in a book of paintings by some of the great artists, like Rembrandt, Monet, Da Vinci, or whoever you like. How do they use light? How does the light enhance the image? Study the lighting and how the artist uses light to convey an idea or emotion. Now think about how you could change the light to create another effect or emotion. Is there something you might call a key light, a fill light, or a backlight? Study the light.

PRODUCTION AUDIO

Introduction

Not long ago I received a DVD from a young man who applied for a job. On the DVD, there was a ten-minute video he made. I sat down with couple of co-workers to see what he had done. When the video was done, the first thing my co-worker said was, "What was up with the audio?" Nobody said anything about the nice camera work, the lighting, or the clever story, but focused on the less than professional audio. Unfortunately even the best video images in the world quickly can be undone by bad audio.

For some reason, we seem to be more forgiving of bad video than we are of bad audio. I have shot video in the past that was poorly lit, poorly composed, and was just plain awful, but the only comment I hear is "What is wrong with the audio?"

How Sound Is Created and How Sound Travels

The sound track for any video includes music, speech, ambient (background) sound, and special effects. Music and special effects are added during the post-production phase, so this lesson deals with speech and ambient sound which you acquire during production.

Before getting into audio for video it is a good idea to understand the basic principles of how sound is created.

Think of throwing a pebble into a lake. The wave starts in the center where the rock displaces the water molecules. The displaced water molecules push against the water molecules next to them and those molecules push against the molecules next to them and so on. Pretty soon you have a wave moving across the lake. A larger rock displaces more water and creates a bigger wave that travels farther.

Vibrating objects create sound waves. When a guitar string is plucked the vibrations disturb the air molecules around it. This starts a chain reaction of air molecules disturbing the air molecules around them. The resulting sound wave travels through the air.

Your eardrums vibrate when the sound wave reaches your ears. The vibrations are then transferred to your brain, which translates the vibrations into sounds we recognize, like a G note played on a guitar (see Figure 7-1).

FIGURE 7-1
How sound travels

Vibrating Guitar strings

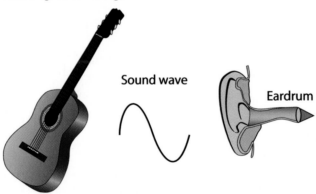

A sound wave has three important properties:

- Wavelength
- Frequency
- Amplitude

Wavelength is the distance between equivalent points on consecutive phases of a wave pattern. It is the length of the wave as shown in Figure 7-2.

FIGURE 7-2
Diagram of a wavelength

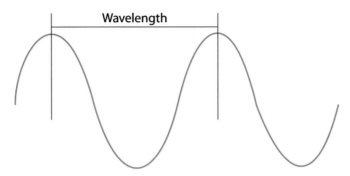

Frequency is the number of times a wavelength repeats in one second as shown in Figure 7-3. A higher frequency means the sound source is vibrating faster and creating more wavelengths per second resulting in a higher pitched sound. A lower frequency means the sound source is vibrating slowly and creating fewer wavelengths per second which results in a lower pitched sound. Frequency is measured in hertz, which you usually see abbreviated as Hz. Figure 7-3 has a frequency of 4 Hz.

FIGURE 7-3
Diagram showing frequency

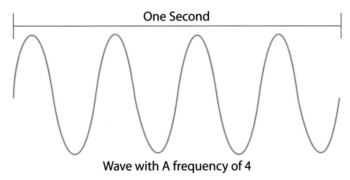

Humans are sensitive to frequencies between 20 Hz and 20,000 Hz, while dogs can hear from around 50 Hz to around 145,000 Hz. This means human can hear lower pitches that dogs cannot hear, but dogs can hear much higher pitched sounds.

Amplitude is the power of the wave. A sound wave with higher amplitude produces a louder sound, while lower amplitude produces a quieter sound. Amplitude is usually referred to as the volume. The volume is measured in decibels, which usually is displayed as dB. In Figure 7-4, the sound wave on the left has a higher amplitude, meaning it would be louder than the sound wave on the right. Both waves have the same frequency, meaning the same sound, it is just that the wave on the left would be louder than the wave on the right.

FIGURE 7-4
Diagram showing amplitude

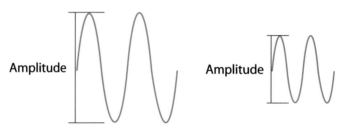

How Digital Audio Is Created

Converting analog audio to digital audio is basically the same process as converting analog video to digital video. Digital audio copies the analog signal by picking points along the analog wave each second and basically assigning each point a numerical value that a computer can work with. This process is called sampling (see Figure 7-5). The idea of digital audio is to recreate the audio wave as closely as possible. The better the digital reproduction, the closer it sounds to the original analog sound. The easiest way to explain this concept is to show it visually, as seen in Figure 7-5.

FIGURE 7-5
An example of sampling

Two basic factors determine the audio quality: sample rate and bit depth.

Sample rate is the number of times in a second that a sample is chosen from the analog signal. CDs, for example, have a sample rate of 44.1 kHz. That means that 44,100 samples of the audio signal are chosen each second (the k stands for thousands). You may think that the higher the sample rate, then, the higher the audio quality; however, eventually the man made replication of the audio cannot get any better, no matter how many samples are taken each second. 44.1 kHz works well and gives high enough quality audio for most people.

Figure 7-6 shows the basic principle of sample rate. If the wave on the left is one second of audio and five samples are taken each second, the resulting digital copy of the analog wave is shown on the right. The digital copy in Figure 7-6 gives a basic idea of what the original sound wave looks like, but it isn't really close. The resulting audio does not sound very close to the original because there are not enough samples to accurately re-create the original wave. That's why the digital copy looks like a city skyline and not a wave.

FIGURE 7-6
Basic principle of sample rate

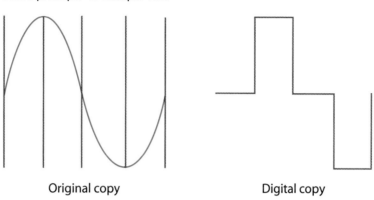

Original copy   Digital copy

Compare the resulting digital copy with five samples shown in Figure 7-6 to the one in Figure 7-7 that shows ten samples each second. A digital wave using ten samples is a more accurate copy than one with five samples, but it is not a very good copy because there still is not enough samples to create an accurate copy of the original wave. If you double that again, you have an even more accurate copy as shown in Figure 7-8. Now imagine a copy of the wave using 44,100 samples instead of only 5, 10, or 20 samples. The resulting digital wave would be very close to the original wave.

FIGURE 7-7
Example of a sample rate of 10 samples per second

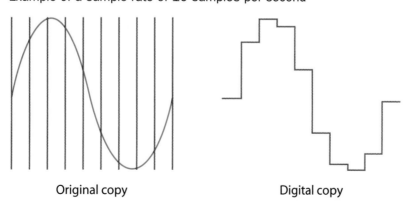

Original copy   Digital copy

FIGURE 7-8
Example of a sample rate of 20 samples per second

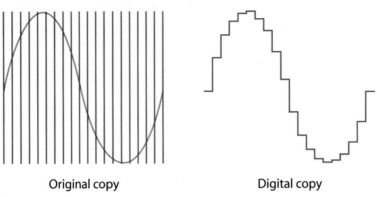

Original copy   Digital copy

While sample rate deals with how often a sample is taken (time), **bit depth** is the number of levels (or how high and how low) a sample can recreate. If you placed bit depth and sample rate on an XY axis, it looks like the diagram in Figure 7-9.

FIGURE 7-9
Bit depth of 4 bits

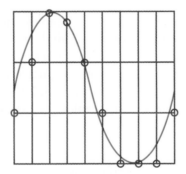

Figures 7-9 and 7-10 show the difference between a bit depth of 4 bit and 8 bit. The dots represent when the sample is taken and how close the bit depth, represented by the horizontal grid marks, allows the sample to approach the original signal. If there is not enough bit depth the sample has to get as close to the original signal as it can as shown in Figure 7-10.

FIGURE 7-10
Bit depth of 8 bits

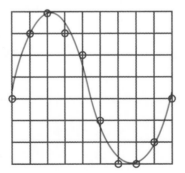

Because of the limited bit depth, the first point taken in Figure 7-9 isn't very close to the wave, neither is the second. The third, fourth, fifth, sixth, and eighth points are pretty close to the wave, but the seventh, ninth, and tenth points are not. That means that only half of the sample points are accurate. The closer the sample points are, the more accurate the resulting digital copy. In Figure 7-10, 7 of the 10 sample points are close to the wave that means the resulting digital copy is a closer copy.

Bit depth and sampling rate determines how much data is needed each second to transfer the data and how large the audio file will be. If you have too little information the audio does not sound very good at all. DV audio is 16 bit, 44.1kHz audio, which is the same as audio CDs.

Digitized audio relies on compression just as digitized video relies on compression. Compression, if you remember, takes all of the information needed for an audio (or video) file and makes it smaller. This allows higher quality audio to travel the same places as lower quality audio.

How Microphones Work

I recently videotaped my son in a school performance. I was late and had to record the first half of the program from the back of the auditorium. When the program stopped for a short intermission, I moved to the front of the auditorium.

> **Note**
>
> A 4 bit sample would actually not be just four levels, but 2 to the power of 4, or 2x2x2x2, or 16 levels. This is because computer languages are binary, which means the only two numbers are 1's and 0's. An 8 bit sample is 2 (which represents the 2 numbers in computer languages) to the power of 8, or 2x2x2x2x2x2x2x2, or 256 levels. A 16 bit sample would be 2x2x2x2x2x2x2x2x2x2x2x2x 2x2x2 or 65,536 levels and a 24 bit sample would be 2x2x2x2x2x2x2x2x2x2x2x2x 2x2x2x2x2x2x2x2x2x2x2x2 or 16,777,216.

> **Note**
>
> The microphones you use, even the expensive, professional models, are analog. The signal comes in analog and then the microphone captures an analog signal. The analog signal is then converted into a digital signal by the camera or digital input. Keep this in mind as we continue this lesson.

When watching the tape later I could barely hear the kids singing at the beginning of the program. Mostly I picked up people talking, kids screaming, and a lot of coughing. The audio for the second half of the program was fine because the microphone was close enough to the stage that it did not pick up most of the background noise.

The goal for recording live production audio is to record only the audio you really want to use. This is done by placing the microphone as close to the subject as possible. That is not to say that the only thing the audience hears is the voices of two people talking in the middle of a crowded room; it would be very unnatural.

Much of what the audio is recording is determined by which type of microphone you use and how you use it. The first thing we look at is how different microphones collect and recreate audio signals. The second thing we look at is the pick up pattern. Lastly, we'll look at how the microphone is going to be used.

All microphones work in the same basic way. Each one has an object or device that vibrates when audio waves reach it and a method for translating those vibrations into an electronic signal. The electronic signal is then moved through a wire and stored on some type of recording media, like a tape or hard drive.

Different Types of Microphones

Although there are several technologies used in microphones, you learn about the three most commonly used technologies:

- Dynamic
- Condenser
- Ribbon

Dynamic Microphone

A dynamic microphone contains a diaphragm that vibrates with the audio waves. The vibrating diaphragm moves a magnet or coil that creates an electrical current that travels down a wire as shown in Figure 7-11. The current is then stored as audio on a hard drive or a tape.

FIGURE 7-11
Diagram of how a dynamic microphone works

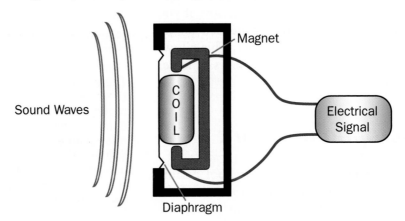

The advantages of using a dynamic microphone are that they are rugged, they do not require a power source, they are also not as expensive as some other types of microphones, and they are not as sensitive to extra movement and handling. The disadvantage is that they do not pick up as wide of a range of frequencies as other microphones.

Condenser Microphone

A condenser microphone uses two metallic diaphragms mounted very closely to a back plate. The diaphragms and back plate are connected to a battery that creates an electrical charge between them. When the diaphragms vibrate, the distance between the diaphragms and back plate change. This change in distance causes a current to flow through the wire as the battery maintains the current between the diaphragm and the back plate as shown in Figure 7-12. The charge is small and needs to be amplified, so a small amplifier is included in the microphone.

FIGURE 7-12
Diagram of how a condenser microphone works

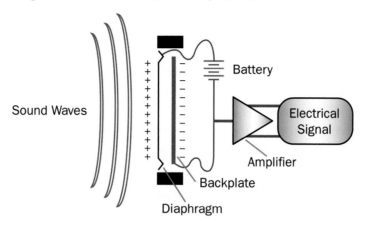

Condenser microphones usually are found in studio settings. The advantage of condenser microphones is that they are more sensitive to a wider range of frequencies, which means a more faithful reproduction of the original signal. The disadvantages are that they require a power source (usually a battery, which really is not a big deal unless the battery dies), and that they are not as rugged as dynamic microphones.

Ribbon Microphones

Dynamic and condenser microphones are more common, but you may come across a ribbon microphone now and then. In a ribbon microphone a metal, corrugated ribbon is placed between the opposite poles of a magnet. The ribbon vibrates and creates a current as shown in Figure 7-13.

FIGURE 7-13
Diagram of how a ribbon microphone works

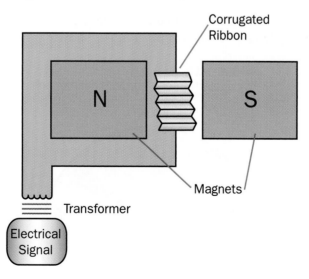

The advantage of a ribbon microphone is they offer high quality sound reproduction, but the disadvantage is they are fragile and breathing too hard on a ribbon microphone can damage it.

Describe Microphone Pickup Patterns

A pickup pattern basically determines the area around the microphone where the microphone is sensitive to sounds. This does not mean that all sounds outside of the pick up pattern completely are eliminated—if a big truck drives by you will record that along with what you really wanted the audience to hear. A pickup pattern does, however, reduce what the microphone picks up and gives you more control over the audio.

An omnidirectional microphone is sensitive to sounds that come from all around the microphone. This means that if you put a microphone in the middle of a crowd the microphone will pick up audio from all directions around the microphone as shown in Figure 7-14.

FIGURE 7-14
Pick up pattern of an omnidirectional microphone

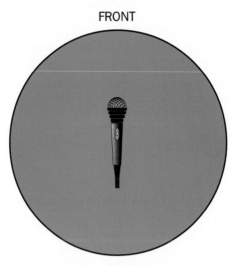

A cardioid microphone picks up sounds in a heart-shape pattern. This means the microphone will pick up sounds from in front, but not from behind as shown in Figure 7-15.

FIGURE 7-15
Pick-up pattern of a cardioid microphone

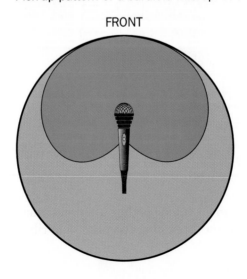

A supercardioid microphone has a narrower pickup pattern in the front than a cardioid and picks up some sound from behind the microphone as shown in Figure 7-16.

FIGURE 7-16
Pickup pattern of a supercardioid microphone

FRONT

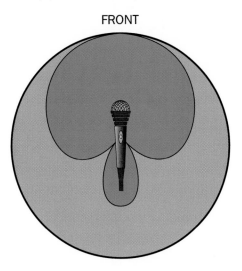

A hypercardioid microphone is similar to the supercardioid, but has an even narrower pickup pattern in the front and a little more in the rear as shown in Figure 7-17. This means it eliminates even more audio from the sides of the subject.

FIGURE 7-17
Pickup pattern of a hypercardioid microphone

FRONT

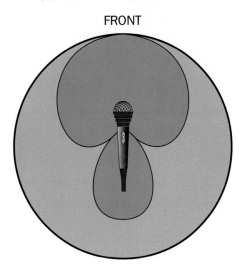

Bidirectional microphones pickup audio in the front and in the rear, but not from the sides as shown in Figure 7-18.

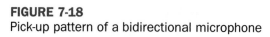

FIGURE 7-18
Pick-up pattern of a bidirectional microphone

FRONT

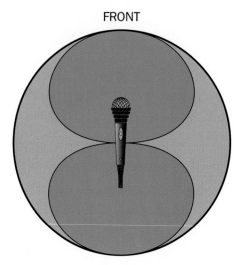

There are other patterns, but these five are the basic patterns that you need to understand.

The Different Ways Microphones Are Used

A video shoot should be as carefully planned out for audio as it is for video. A scene from an action film would not work very well if the hero had to hold a microphone in one hand while he was trying to climb a cliff to save someone who is trapped. We do expect a reporter on a breaking news story, however, to stand in front of the camera with a microphone.

The camera-mounted microphone is the first microphone we talk about, but it should not be your first choice when selecting a microphone. The camera-mounted microphone is usually some type of supercardioid or hypercardioid microphone that allows parents to record their children's school plays and programs from anywhere in the room. Remember, however, that the idea is to get the microphone as close to the subject as possible to reduce unwanted sound pickup. The camera-mounted microphone is good for picking up ambient, or background, sound.

The second type is hand held microphones. You usually see hand held microphones used by reporters on television or by singers in concerts. I also have heard these microphones referred to as stick mics. The advantages of hand held mics are that they can be positioned wherever they need to be. They can be placed closer to the speaker's mouth if there is a lot of extra noise, or quickly moved to somebody being interviewed. These microphones work great if you do not care if the audience sees the microphone.

The third type is boom microphones. **Boom microphones (mics)** are microphones that are connected to a boom pole (often called fish poles). Boom mics usually have a supercardioid or hypercardioid pickup pattern that allows the audio operator to eliminate almost any unwanted noise. A boom mic should be as close to the subject as possible without being seen by the audience. A boom mic above the subject's head (and out of frame) is the first choice of most audio operators with whom I have worked. A single boom mic can be used to pick up audio from a number of subjects without a great deal of problems. The disadvantage of a boom mic is that you need somebody to hold it. I have seen cradles that allow you to connect the boom pole to a c-stand, but you have to make sure the boom is where you want it before you start shooting. When using a boom mic make sure its shadow does not appear in the frame or that the microphone does not drop into the frame.

The fourth type is **lavalier microphones**. Lavalier (or lav) microphones are small microphones that can be attached directly to the subject's clothes. (It is not uncommon to tape a lav mic onto a person's skin.) The lav should be placed about six to eight inches from the speaker's mouth. One of the trickiest parts of placing a lav mic is the cable. The cable should run under a shirt or blouse so it is not seen, or even run down a pant leg if the subject is seen in a long or full body shot.

I use lavs a lot in my job, but mostly because I do not have anybody to hold a boom for me. Lavs do a good job picking up audio; they are easy to use (especially for a one person operation) and with some practice the audience does not see the cables. The problem with lav mics is their strength and their size. Because lav mics are so small it is easy to break the small wires. Other problems are that they pick up the sound when the mic rubs against clothes, and people often forget they have them on and either hit them while they are speaking or run off with the microphone still attached to the recording device.

Wired Systems and Different Audio Connectors

It seems like every time somebody asks me to use a microphone they ask for a wireless. Wireless has that wow factor, but wired microphones are actually more reliable and create higher quality audio. With that in mind wireless microphones do have their place.

A wired microphone has wires that connect the microphone to whatever recording device you are using. Audio uses a number of different connections, but the connectors you are most likely to use are shown in Figure 7-19. They are:

- RCA
- Mini
- XLR
- Telephone or TRS (tip, ring, and sleeve)

FIGURE 7-19
Different audio connectors

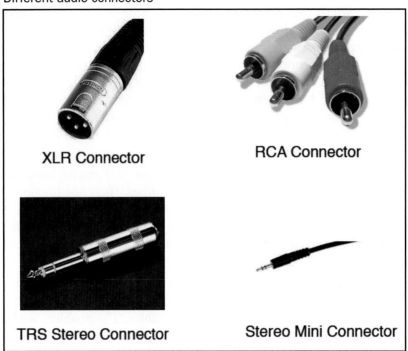

XLR Connector

RCA Connector

TRS Stereo Connector

Stereo Mini Connector

The most common audio inputs and outputs on consumer or prosumer DV cameras are RCA connectors and mini connectors. I have not seen any microphones with RCA connections, but there are many different microphones, from handheld to lavalier, that use mini connectors.

Audio using RCA and mini connectors is called unbalanced audio. Unbalanced audio cables and connectors use two lines to transmit the signal, which makes them susceptible to noise like hums and hisses from electronic equipment.

Balanced audio lines and connectors, such as XLR and tip, ring, and sleeve telephone (TRS), use a third cable that helps reduce the hum and hiss. Balanced audio provides a cleaner signal and allows the signal to move over greater distances. Greater distance allows you to get the shots you want from whatever distance or position without compromising audio quality. Just because many consumer cameras do not have balanced inputs does not mean you cannot use high quality, balanced microphones. Line balancers and adapters allow you to connect high quality microphones through unbalanced audio connectors.

Wireless Systems

Wired and wireless systems are very similar; however, wireless microphones differ in the way they transfer the audio signal to the recording device. A microphone is connected to a wireless transmitter that can be attached to the subject. A receiver is then connected to the camera or recording device.

Note

The RCA and mini connectors used in video cameras are analog inputs and outputs. The connectors on the camera allow you to use your camera to watch the tape on a regular television. The digital output, if you'll remember back to Lesson 1, is the Firewire or DV connector.

Other Audio Equipment

One of the nice things about shooting with digital video is that you can record the audio directly to the camera. Just connect the microphones directly to the camera and go. Shooting with film gets a little more complicated because you do not record the audio directly to the film. This means you do not need a whole lot of other equipment; however, may I suggest a few other pieces of equipment that make your life easier and help you record better audio.

The first piece of extra equipment you will want is a good pair of headphones. Headphones let you hear what you are recording. If, for example, a truck drives by while you are shooting, you will know whether or not you will have to re-shoot so you won't hear the truck.

Note

Not all transmitters and receivers are created equal. A high quality wireless system that gets close to the audio quality of a wired microphone can cost thousands of dollars. While the audio from a wireless system is not as good as a wired system, most people can't really tell the difference when a good system is used.

Headphones also allow you to check the audio levels. The audio level measures the volume of the audio coming into the recording device. Most cameras give a visual representation of what that level, or volume, is. This volume is displayed in dBs or decibels. The lowest level that humans can hear is approximately 0dB. If the audio is too loud, the signal is too high and distorted.

Even though you have meters and measurement tools for audio, listening is still one of the best ways to know how high the levels are. If the levels are too low or too high you need to know how to adjust them. Most of the cameras I use have simple dials I can turn to adjust the levels higher or lower. One camera, however, has a menu that I have to go into in order to adjust the levels. Many consumer and prosumer cameras have the same type of menu-adjusted levels. Make sure you understand how the camera you are using works.

The second piece of equipment I suggest is a field mixer. A field mixer is a portable audio mixer that runs on batteries so you can adjust the microphone audio levels. The microphone is plugged into the mixer and the mixer is then connected to the camera. Most cameras have only one or two audio inputs, but a mixer allows you more, depending on which one you use. A mixer also gives you more control over the input signal, especially if the camera doesn't offer much audio control.

Note

Analog and digital video can get confusing and complicated. Analog audio equipment should be set at 0dB, while digital equipment shows that same level as -12 dB. When you are working with your equipment make sure you read the owner's manual and understand exactly how your equipment works and where things need to be set.

Some field mixers also include a tone generator. Have you ever stayed up very late watching TV and seen the color bars on the television along with that high-pitched tone? That tone is known as a 1K tone. The 1K tone is set at a consistent level and helps you set recording or playback audio levels at the same position so your audio sounds right.

A final piece of equipment you might want to consider is a small back-up recorder. When I have an important shoot I like to use a portable digital recorder to make sure I have the audio I need. When I go into post production I capture the audio from the back up recorder if I need to and synchronize it with the video when I edit.

Keys to Acquiring Good Audio

So far we have covered how sound is created, how microphones work, as well as other equipment you might want to use. Getting good audio has less to do with equipment and more to do with people.

The first key to getting good audio is to scout the location with audio in mind. Not long ago I had to scout a location, but I did not do it right—I was more concerned with the video than I was with the audio. I ended up with good video, but the bad audio almost ruined the shoot.

Make sure you scout the location at the same time as when you will shoot. If you are scheduled to shoot on Thursday at nine, for example, make sure you visit the location on Thursday at nine. If you have the time and the patience, visit the location a few times before you shoot to make sure you have a good idea of what is going on.

Also, make sure you close your eyes and listen. When I scouted the previous location I just talked about, I listened with my eyes open. When I shot the location, I heard the same sounds as when I scouted, but the sounds seemed to be much louder when I was shooting. Why? I concentrated more on the audio while I was shooting than I did when I was scouting. Closing your eyes and focusing solely on what you hear helps you get a much better idea of what is likely to happen while you are shooting.

Of course you do not always have the chance to scout locations when you are reporting the news. Asking people to be quiet, turn off their car, or get their dog to be quiet is not as hard as it may seem sometimes—you would be surprised how cooperative people can be when they know you are shooting video.

The second key is to have people that concentrate on audio only. This can be a problem on small shoots when you need people to do whatever it takes to get the job done. Having at least one person concentrating on audio, however, really helps. The audio person should listen to the audio to check levels as well as make sure there is no unwanted noise. If there is a problem the audio person needs to let the director know and quickly figure out how to solve it. There is nothing worse than shooting and finding out the audio was bad after everybody already has gone home.

SUMMARY

In this lesson you learned:

- Sound is created by vibrating objects. The vibrations disturb the air molecules around the object, which in turn disturbs the air molecules next to them. This creates an audio wave that travels through the air.

- The wave attributes that are important to understand are wavelength, frequency, and amplitude.

- Frequency is measured in hertz (Hz), while amplitude is measured in decibels (dB).

- Digital audio is created by sampling, much the same way as digital video is created from analog video. Bit depth determines how close each sample can get to the original signal. DV audio is 16 bit 44.1 kHz.

- All microphones work in the same basic way. First, an element, usually a diaphragm, vibrates with the audio waves. The vibrations create an electrical current that is sent to some type of a storage device.

- The three most common types of microphones are dynamic, condenser, and ribbon.

- Each microphone has an area around it that is sensitive to sound called a pickup pattern.

- Microphones can be camera mounted, handheld, boom, or lavalier.

- The four commonly used audio connectors are RCA, mini, XLR, and telephone. RCA and mini are unbalanced, while XLR and telephone connectors are balanced.

- Balanced audio uses a third wire that helps eliminate noise and hum.

- Wireless systems consist of a transmitter connected to the microphone and a receiver connected to the camera.

- Headphones, field mixers, and portable recorders are other pieces of equipment you might want to have.

- People are the key to acquiring good audio. Scouting locations at the same time that a shoot is scheduled, listening while your eyes are closed, and having somebody focusing solely on audio helps make sure that the audio is as good as possible.

VOCABULARY *Review*

Define the following terms:

amplitude	dynamic microphone	pick up pattern
balanced audio	field mixer	RCA connector
bidirectional microphone	frequency	ribbon microphone
bit depth	hypercardioid microphone	sample rate
boom microphone	lavalier microphone	supercardioid microphone
cardioid microphone	omnidirectional microphone	telephone or TRS connecter
condenser microphone	mini connector	wavelength

REVIEW *Questions*

TRUE/FALSE

Circle T if the statement is true or F if the statement is false.

T   F   1. The sound track includes music, speech, ambient sounds, and special effects.

T   F   2. Telephone connectors are unbalanced.

T   F   3. The camera-mounted microphone is the best choice for all situations.

T   F   4. Amplitude is measured in Hertz.

T   F   5. Vibrating objects create sound.

T   F   6. Amplitude refers to the wave power.

T   F   7. Lavalier microphones are always wireless.

T   F   8. RCA connectors are unbalanced.

MULTIPLE CHOICE

Circle the correct answer.

1. What is DV audio?
   A. 24 bit 44.1 kHz
   B. 16 bit 48 kHz
   C. 16 bit 44.1 kHZ
   D. 24 bit 48 kHz

2. What does 44.1 kHz mean?
   A. 44 samples are taken every second
   B. 44 levels of compression
   C. A type of audio compressor
   D. 44,100 samples each second

3. Which type of microphone requires battery power?
   A. ribbon
   B. dynamic
   C. condenser
   D. crystal

4. Which option is true about balanced audio?
   A. Means using two microphones
   B. Is another description for stereo
   C. Removes audio noise
   D. Equal parts music and dialogue

5. What option applies to ribbon microphones?
   A. Uses a wire suspended between opposite poles of a magnet
   B. Are best for location audio
   C. Is another name for a boom microphone
   D. Are always bidirectional

6. Where does an omnidirectional microphone pick up audio?
   A. in front of the microphone only
   B. from the sides of the microphone only
   C. from the front and the back of the microphone
   D. all around the microphone

7. What is a decibel?
   A. Measurement of the power of a sound wave
   B. The most important wave property
   C. An omnidirectional microphone
   D. Measurement of the number samples taken

WRITTEN QUESTIONS

1. Describe sampling.

2. Explain the basic idea of how a microphone works.

3. How should you scout locations with audio in mind?

4. Explain how planning for audio helps determine what audio needs you have.

PROJECTS

PROJECT 7-1

Experiment with any microphones you have and figure out how the pickup pattern for each microphone affects the recording. If you are not sure what the pattern is, read the manual or experiment to figure out what it is. Try to decide for which situations the pick up pattern would be used. How close does the microphone need to be to the subject to do a good job picking up the audio signal?

PROJECT 7-2

Scout a location (whether you are planning to shoot or not). Scout on two different days of the week at two different times of day. Identify the different sounds and sound intensities between the two times. Is there a difference in what you hear? Scout with your eyes open, paying attention to what you are looking at the same time you are listening. Now do it with your eyes closed. Is there a difference in what you are hearing. Write down what you hear and observe.

PROJECT 7-3

Plan a scene for audio. What type of microphone will you use? Where is the best location for the microphone? Will the microphone be a boom mic? A lavalier mic? What other type of equipment will you need? What crew will you need? Describe the scene in writing and include the answers to the questions listed in this Project.

WEB PROJECT

Research microphone manufacturers on the Internet. Identify which pickup patterns are used for which microphone. Compare the specifications (such as frequency amplitude) for different microphones and identify what that could mean for your use.

TEAMWORK PROJECT

Get together in teams and shoot a scene or a newscast. Each person should take a different job, such as being in the scene, working on a mixer (if you have one), holding the microphone, or whatever. You don't have to shoot video, but it is a pretty good idea. Once the shoot is completed, listen to the audio and identify areas in which the audio can be improved, such as different microphone placement, or a higher recording level. Re-shoot or record, and repeat until the audio sounds the way you want it.

CRITICAL *Thinking*

ACTIVITY 7-1

Watch a film, video, or television show, and identify the audio elements you think were acquired during production and those that were added during post-production. What could have been done to improve those elements acquired during production? What effect does the elements added during post-production have on the elements acquired during shooting? How does the audio as a whole add to or detract from the final production?

DIRECTING

Introduction

In this lesson you learned how the video camera works, how to use a camera, and how to record good audio. You could now go out and shoot your fictional narrative, documentary, or news script, and be okay; in this lesson, we cover directing before we move on to editing video.

Director's Responsibilities

A few years ago I worked with a guy who had been in the movie industry for years as a grip. He took great joy in making fun of directors and told me that directing "is so easy anybody can do it." He was right in one way. Anybody can be a director, but it takes hard work, talent, and an understanding of how to tell a story to be a good director.

When we think about the director we usually think about somebody telling the actors what to do, and then screaming "Action!" and "Cut!" If directing were that easy anybody that could scream could be a director.

The script gives the director the blueprint, but what the audience hears and sees is the director's interpretation of what is on the page. The director may even choose to (and often does) re-write the script. After the script is finished, the director approves everything, from costumes to makeup to props to lighting. The director needs to help cast actors or on camera talent, pick the crew, and then make sure everybody does their job. Then, when everything is shot, he or she sits down with the editor to put everything together.

For most shoots the director's success is determined in the pre-production stage. If the director knows the script and the story, and has a vision of how to tell the story, then the video has a good chance of being successful.

The key for making the pre-production phase successful is to put everything on paper. Do not think that you will remember every brilliant idea you have if you do not put it down on paper. If you have an idea about how a scene will play out, write it down. If you have an idea about a certain shot, draw a picture. The picture does not have to be a work of art, but it should help you visualize what happens.

Even with all of your ideas and planning you may find a better shot when you are setting up. You can make the change because you had a starting point. There is nothing worse than a director that can not make up his mind and shoots everything from every possible angle because he or she is not sure what the scene will look like. Most movies and documentaries you see are shot using a single camera, while most sporting events and television programs, like talk shows, are shot using multiple cameras.

The director's most important job, however, is to tell the story. All of the special effects, beautiful people, pounding soundtracks, and plot twists in the world can not save a story from a director that does not know how to tell one.

Shooting to Edit

Not long ago a student came to me with a project he was having problems with. He had a lot of footage, but when he tried to edit everything together, nothing looked like it fit together. Everything was shot on the same day in a short amount of time, but the characters or action seemed to jump with each edit. Unfortunately the only solution was to re-shoot each of the three scenes.

It is the director's responsibility to make sure the footage fits together to tell the story. This is called shooting to edit. It is easy to make sure the footage fits together when you follow a few simple guidelines concerning shots and sequences.

A shot is the time between when you start recording and when you stop. It is the basic building block for video. If you record for ten seconds and stop recording, you have a 10 second shot.

Each shot should have a purpose and convey meaning. Three ways that meaning can be expressed are camera position, composition, and lighting. These three ways of conveying meaning are concepts you already know and understand from earlier lessons. The director works with the cast and crew to setup each shot to match the director's vision. The cast and crew may have suggestions that the director might want to consider, but the director needs to make the final decision.

Building a Sequence

Let's say you have a scene where two people are talking. The dialogue is going to last about 25 seconds. What is the best way to shoot the scene? The easiest way is to set the camera up so the audience can see both people at the same time. The camera does not move and the people deliver their lines. This is a one shot scene.

Look at Figure 8-1 and count how many seconds you can look at it before your eye starts to wander away.

FIGURE 8-1
One shot scene

How did you do? I had to force myself to look for 10 seconds, but I wanted to look away sooner. It does not really take long to gather in everything the image has to offer. The audience probably will not be interested in your video for very long if each scene consisted of a single shot.

A second way is to set the camera up and pan back and forth between the two characters. This way would work, but it is still a single shot and the audience may get seasick.

A third way is to shoot a number of different shots, then edit them together. This is a better idea, but what shots do you use? A bunch of long shots? A bunch of close-ups? A mixture of long shots and close-ups? A series of individual shots edited together to create a scene is called a sequence. There are all sorts of ways to build a sequence, but, like story structure, there is a basic, classic starting point.

The first shot is a wide shot commonly called the **establishing** or **master shot**. The master shot establishes where the character is and what is going on around them. You know you have a master shot if you could show everything that happens in the scene without using another shot (Figure 8-2).

FIGURE 8-2
Master shot

The master shot of a girl playing basketball might be a wide shot of a girl playing basketball on a basketball court. The arena or the bleachers do not need to be included unless they are important to what is going on. In the case of a conversation (shown in Figure 8-2) the establishing shot is of a man and a woman sitting in a diner. We do not need to see anybody else in the establishing shot, unless it is important that the audience knows that other people are there, like if the script calls for the scene to happen in a busy diner.

The second shot would be a close-up or medium shot. If character A is talking, the close-up or medium shot is of her (see Figure 8-3). When character A is finished talking, the next shot is of character B (see Figure 8-4). The rest of the conversation is alternating close-ups of character A and character B (see Figure 8-5).

FIGURE 8-3
Close-up of character A

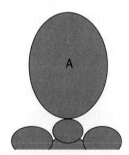

FIGURE 8-4
Close-up of character B

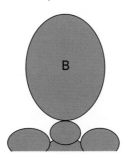

FIGURE 8-5
Alternating close-ups between characters A and B

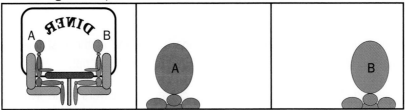

Using close-ups does a couple of things; first, it draws the audience to the character. A long shot does not allow the audience to see the character's facial expressions or be involved in what the character is thinking and feeling. Second, it gives the audience something new to look at. The worst thing that can happen is that the audience looks away from the screen. That is a basic sequence: a master shot, a close-up, and a close-up, but the sequence easily can be improved.

One way to improve the sequence is by adding reaction shots. A **reaction shot** shows the listener's reaction to what the speaker is saying. Reactions shots create relationships between characters. If the woman tells the man she never wants to see him again and he cries, the audience understands he is hurt. If he shrugs his shoulders then the audience understands he does not care. A good director uses those shots to create relationships, express character, develop story, and create tension or drama. A reaction shot is a close-up of a character that is not speaking.

The last thing we can add to the sequence to add interest and flavor is a cutaway. A **cutaway** is a shot of something extra in the scene. For example, one of the characters pouring sugar into their drink in our diner sequence is a cutaway. A cutaway in an interview for a documentary is a graphic of whatever the person being interviewed is talking about.

That is all that is involved in a basic sequence. You find that most scenes and sequences follow this structure. Even the news is shot using this basic sequence: open with a shot of the two anchors, then a close-up of the anchor with the lead story.

You can change the structure, but you better have a good reason to change it. Say you had a character that ends a scene by saying they would never, ever, in a million years, eat a bowl of corn flakes. The first shot of the next scene might be a close-up of a bowl of corn flakes, followed by a long shot of the character sitting

> **Note**
>
> This basic sequence is simple, but it can be so powerful. Say, for example, you have a shot of a man saying "I want peace," followed by a shot of a violent explosion. The contrast of what is said and what is shown creates a power emotion. This is a more advanced editing theory, but keep in mind that the basic sequence can be used in a number of ways, so master the basics.

at the table getting ready to eat. The close-up is normally a cutaway, and the long shot is normally the establishing shot. Reversing the order creates tension or humor. Just make sure you know why you are doing it.

STEP-BY-STEP 8.1

In this Step-by-Step you are going to pull out the script you wrote for the earlier lessons. You will read the script and determine at which point the master shot will be shot, and where close-ups will be shot.

1. Print out **SBS2-8.fdr** or **SBS2-8W.fdr**.

2. Identify the master shot. If, for example, it is a conversation between two people at a diner, the master shot would be the opening shot and end after the first line of dialog for character A is finished. Write it on the script.

3. Identify the first close-up. If the master shot ends at the end of the first line of dialog, the first close-up would be when character B starts to talk. Write it on the script.

4. Identify the second close-up. This would be when character A delivers a line. Write it on the script.

5. Identify any cutaways or reaction shots. This might be a shot of when character A reacts to something said by character B, or when character B reaches for the salt. Write it on the script.

6. If you write electronically on the script, save the script as SBS8-1.doc. If you write in hand on the script, turn in the handwritten copy to your instructor.

Defining "The Line"

The establishing shot establishes where things are in the frame, especially in relationship to each other. It is important to maintain those relationships or you confuse the audience.

Let's go back to our conversation in the diner. Character A should be on the left-hand side of the screen looking toward character B for her close-up, while character B should be on the right-hand side of the screen looking toward character A for his close-up. The audience will be confused and unsure of what is happening if character B is suddenly on the left-hand side of the screen looking in the other direction. It would seem that that the characters are jumping all over the place. You do not want to confuse the audience.

In order to avoid confusion, a concept called "the line" can help you. "The line" is an imaginary line drawn right down the middle of the top of the heads of the two characters as seen in Figure 8-6. If the camera is setup at position 1, all other setups should be on the same side of "the line," such as positions 2 and 3. Setting up the camera this way keeps character A on the left and character B on the right in every shot.

FIGURE 8-6
Diagram showing "the line"

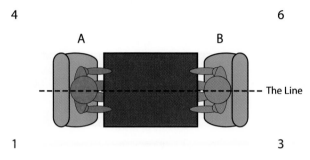

If you setup in positions 4, 5, or 6 the characters switch screen position, with A on the right and B on the left. That is called jumping or crossing "the line." Compare Figure 8-7 with Figure 8-5.

FIGURE 8-7
Jumping or crossing "the line"

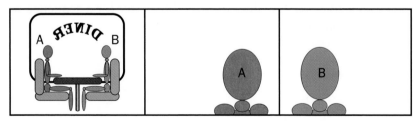

Sometimes, you need to cross "the line" in order to show what is going on. Two ways to cross "the line" is to first use a continuous camera move, or second, cut to a neutral shot.

A continuous camera move allows the audience to see changes in position as they take place, as shown in Figure 8-8. If you decide it is important that the audience sees something behind the camera, such as the counter in the diner scene, the camera has to cross "the line." A continuous camera move, such as a dolly shot, shows that the screen positions are changing and keeps the audience from getting confused.

FIGURE 8-8
Continuous camera move

A neutral shot, or stopping on "the line," basically eliminates "the line." If you have a shot of a car moving from the left to the right then show a shot of the car going from the right to the left it will confuse the audience. If you put a shot of a car coming toward the camera (a neutral shot) between the two shots, however, "the line" does not really exist in the mind of the audience. Using a neutral shot to cross "the line" is probably better for a sequence with a single subject than in a two-person conversation (see Figure 8-9).

FIGURE 8-9
Neutral shot

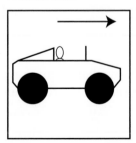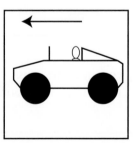

The car is moving to the right. Cut to a neutral shot. Cross the line to cut to the car moving to the left

At other times "the line" changes positions. If character A, for example, gets up and moves to the jukebox behind character B, you have a new orientation in the scene: "the line" has changed positions. When "the line" moves it should happen in the frame so the audience can see the change. For example, a close-up of character B looking to the left followed by a shot of character A at the jukebox will be confusing. If you show character A moving, such as in a long shot, the audience knows what's going on (see Figure 8-10).

FIGURE 8-10
Line changes position shot

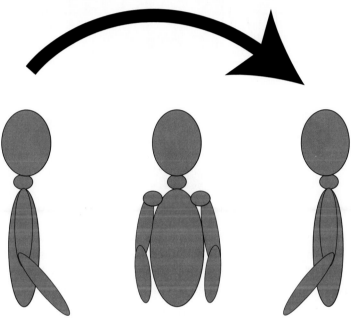

Shooting the Sequence

Most small video projects, as well as most multi-million dollar Hollywood movies, are shot using a single camera, especially for fictional narratives or documentaries. Using one camera means the camera has to be moved and setup a number of times. You could setup for each line of dialog in the diner scene, meaning you setup and shoot the master shot, then re-set and shoot his first line, then re-set and shoot her first line, then re-set and shoot his second line, then re-set and shoot her second line and so on.

Do not do it that way. This diner sequence requires three setups: First, a two shot (both characters in the frame at the same time); second, a close-up of the woman; and third, a close-up of the man (see Figure 8-11). You can shoot individual dialog, reaction shots, and cutaways from the two close-up camera setups.

FIGURE 8-11
Three camera setup

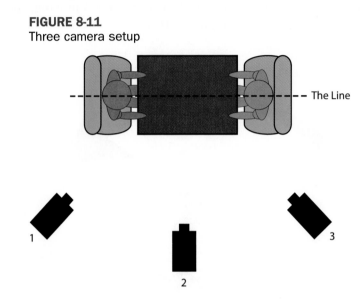

Once you know how many setups the scene takes you can decide how to shoot the sequence. The first way is to shoot the entire scene from camera setup one, then shoot the entire scene from camera setup two, and then shoot the entire scene from camera setup three. Shooting the sequence this way gives you plenty of footage to edit, but it takes longer to shoot and edit.

The second way is to determine which shots you will use from each setup. Suppose there are eight lines of dialog in the scene. After reading the script you decide that the master shot is lines one and two. The third, fifth, and seventh lines is the close-up on the man. The fourth, sixth, and eighth lines is the close-up on the woman. You also shoot reaction shots for lines five and eight because of how important those lines are. The camera setups for each shot are determined by what the scene needs. The problem with shooting this way is you might not get enough shots to choose from when you start to edit.

Make sure you shoot everything you need for whichever method you decide to use. It is a pain to match the look and feel of something you shot earlier. It is helpful to have somebody mark up the script for a documentary or fictional narrative after each scene is shot. I like to draw a big X through the middle of every scene to make sure I do not forget everything.

To make sure each shot in the sequence fits together, no matter how you shoot it, you need to pay close attention to what the actors are doing in each shot. If the actor is looking one direction in one shot and in another direction in the next shot they will not match when put together. For example, if the actor looks to the right in the master shot they need to look to the right in the close-up. If the actor has a hand on their chin in the master shot, the close-up will not work if the actor's hand is on the table. Two shots like that would look odd to the audience. Watching for continuity is making sure that each shot matches the other shots.

Changing Size and Angle for Each Camera Setup

All action in the second shot needs to be in the same general area as it is in the first shot. Notice I said general area, not the exact same area. It would be almost impossible for an actor to repeat the exact same action in the exact same area over and over again. With that in mind, there is a difference in the action, no matter what you do. The best way to make the differences less noticeable is to remember one simple rule: change size and angle.

The master shot is most likely going to be a long shot or an extreme long shot. The close-up should be just that, a close-up. That just takes care of the first part of the rule, which is change size. Changing size helps cover the difference in the motions. If you cut from a long shot to a close-up without changing the angle, it looks like the camera suddenly jumped closer to the subject. That is called a jump cut and it looks like a mistake to the audience.

Changing the angle, as well as image size, smoothes the shot change and eliminates jump cuts. When you shoot single camera, for example, the actor needs to perform the lines and actions for the master shot, and then again for the close-up. It is very difficult to match the exact movement from shot to shot. By changing the size and angle, the actor does not have their head in the exact same position in the close-up as in the master shot. The second setup in the diner sequence (the close-up for the man) needs to be at a different angle than the master shot. The angle should be enough of a change so that the change is clear. If the angle is just a couple of degrees different it is hard for the audience to see the angle change (see Figure 8-12).

FIGURE 8-12
Changing the angle

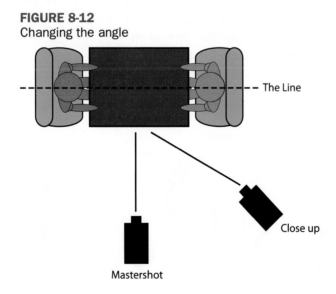

The Line

Close up

Mastershot

Clean In and Out

One last way to shoot so that the edit goes smoothly is called clean in and out. Clean in and out makes cutting on action easier. Say you want to show a person pushing a button with a long shot of the person reaching for the button followed by a close-up of the button being pushed. This requires two setups and it can be difficult to match the action from the two shots.

The easiest way to make sure the shots work together is to shoot a clean in and out. This means that the close-up should show the button in the frame with the hand coming into the frame, pushing the button, and then the hand leaving the frame. Figure 8-13 shows what the shot would look like.

FIGURE 8-13
Clean in and out

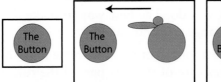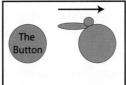

For example you want to show a car driving from a house to the store. The trip may take 15 minutes in real time, but you do not want 15 minutes of somebody driving in traffic. You could compress the time by editing a shot of the car in the neighborhood with a shot of the car in the city, but the change of backgrounds and locations would jar the audience and look like a mistake.

A better way to compress the time and make the sequence look right is to first have the car pull away from the house and out of the frame. Hold the shot of the neighborhood for a second and then cut to a shot of a store. The car would then drive into the shot of the store. The audience has not seen the car for a second or two, so it is easier for them to accept the idea that the car drove from one location to another (see Figure 8-14).

> **Note** ☑️
>
> Clean in and out is not needed in multi-camera shoots because you can use one of the cameras to get a close-up of the button as the hand is reaching for it. The action matches because it is the same action. If you have **matching time code**, meaning the time code for each camera is the same at the same point of the program, you can edit the two shots together easily.

FIGURE 8-14
Compressing time in a shoot

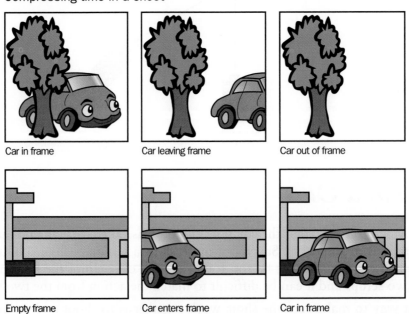

Car in frame   Car leaving frame   Car out of frame

Empty frame   Car enters frame   Car in frame

Setting up for Multiple Cameras

Shooting with multiple cameras should follow the same rules as a single camera shoot. The typical opening shot in a newscast is a shot of the two anchors on the set. This is the master shot. The next shot, after one of the anchors welcomes the audience and presents the first story, is a close-up of one of the anchors telling the first story. The pattern continues throughout the show: master or establishing shot, followed by a close-up.

A multiple camera shoot allows you to setup one time and shoot everything you need without having to reset cameras for every shot. The basic multiple camera setup is three cameras: one camera, we'll call it camera 2, is setup for the master shot, while cameras 1 and 3 are setup for close-ups. This simple setup is common for newscasts, sports shows, talk shows, and presentations. Many multi-camera shoots follow the setup shown in Figure 8-15.

FIGURE 8-15
A multiple camera shoot

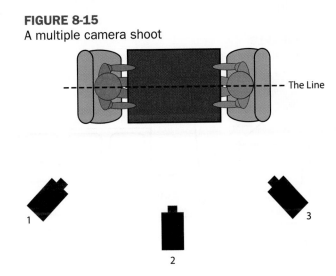

Notice the angle between cameras 1 and 3 and camera 2. Cameras 1 and 3 do not have to be closer to the subjects than camera 2. You can get a close-up by zooming in.

The same principles should be used when setting up for a sporting event: One camera gives you a view of the entire field (the master shot) with the other cameras located at different areas of the field to cover the action and provide options for medium shots and close-ups. The director decides where to set up the cameras long before the shoot. This is determined by location scouting. You also can figure out where to place the cameras by watching nationally broadcast sporting events and trying to figure out where the cameras are set up.

How a Multi-Camera Shoot Works

We have talked about setting up the cameras for a multi-camera shoot, but how does a multi-camera shoot actually work? One way to handle a multi-camera shoot is to have each camera record to a tape. The problem with that is you have to capture all of the footage into a computer and then edit the footage together.

A simpler way to handle a multi-camera shoot is to edit the program together while shooting. The video output from each camera is connected to a switcher. A **switcher** accepts the signal from any number of sources, but allows you to determine which shot is edited into the show, which is called the program. The switcher is then connected to a tape machine (or VTR, video tape recorder). The setup allows the director to see the shot from each camera, as well as the program. This allows the director to determine which camera to switch to and to make sure each camera has a shot that can be used.

A source does not have to be a camera, it can be another tape machine, a graphics machine, a computer, or almost anything that can send a video signal. When a sports highlight is played, for example, a video machine is assigned as a source in the switcher. Figure 8-16 shows a basic setup for a multi-camera shoot.

> **Note**
>
> Switchers come in a number of types, sizes, and prices. I have used simple AV switchers, like the ones you use in a home theater, as a video switcher for a live camera shoot. It is not as clean as a real switcher (and it does not have transition, or effects) but it works. The point is that you can do a multi-camera shoot without spending a lot of money on a switcher.

FIGURE 8-16
Basic setup for a multi-camera shoot

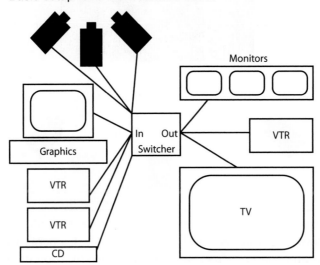

Working with Crew Members to Set Up for a Multi-Camera Shoot

One person can set everything up and shoot using a single camera; however, multi-camera shoots require help.

The most obvious help you need are camera operators. You usually need a camera operator for each camera; however, you can get around that. If the center camera is setup and then locked down (meaning it will not move at all) you can use one less camera operator. The problem with this is if the shot changes, there is nobody there to reframe the shot. Another problem, which is more important, is that the camera could fall. You always should have somebody next to the camera to make sure it does not fall since cameras are expensive and fragile.

The next help you need is an audio operator. You can monitor the audio yourself if you are directing, but it is better if you have somebody that can concentrate specifically on the audio. It is hard to fix any audio problems when you are trying to figure out what shot to use.

You also may want a tape operator. A tape operator starts and stops any recording machines as well as plays back any pre-recorded tapes, such as highlights. The director could be the tape operator, but if the director is monitoring audio at the same time as he is trying to direct and run the tapes he is not going to be very effective or happy.

A common help, but not always necessary, is a graphics operator. A **graphics operator** creates the graphics, runs the graphics computer, and makes sure whatever graphic is needed is ready to be shown at the right time. Often graphics are referred to as a font since you use fonts to create names and text graphics.

The last help you probably want is a technical director. The **technical director** (TD) runs the switcher.

> **Note**
>
> The director needs to be able to communicate with each member of the crew in order to direct the shoot. Communicating before the shoot starts is important, but a director needs to talk to each member of the crew during the shoot as well. This can be hard for a couple or reasons. First, the director and switcher are generally in a different room from where the shoot is taking place. Second, if anybody talks, including crew members, it is recorded. Some type of headset is needed. Professional broadcast cameras have a headset based communication system built in, but you can use walkie-talkies with headset microphones just as well. Just make sure everybody is on the same channel.

Directing a Multi-Camera Shoot

Each camera operator needs to be able to frame each shot well, but the director has the final say on each shot. If the director does not like a shot, they need to tell the camera operator what to do. If the shot in camera 1 needs more nose room the director needs to say something. If camera 3 needs to get a two shot of the two anchors the director needs to tell the camera operators.

When deciding what camera shot to use, the director needs to communicate what he or she wants with both the technical director and different crew members. If the director wants to use the shot from camera 3 they need to say "ready three," which really means "I'm going to use the shot you have now, so do not move." If the camera operator does not know their shot is coming up they may change the shot just as the director goes to their camera. When the director wants to switch to camera 3 they say "take three." The technical director then pushes the button that switches the shot from camera 3 into the program. Switching to a tape or a graphic is handled the same way with the director saying "ready to roll tape" or "ready font," followed by "take tape" or "take font."

It is pretty simple in theory, but the director needs to know what each camera is doing, which graphic or font is needed next, and which tape is in which machine to play back. Even with all the pre-production and planning, things can still go wrong, but the better prepared you are, the smoother things go.

So how long should each shot last? There is no formula. You have to figure out how long each shot needs to be while you are editing or switching, just make sure you get enough coverage while you are shooting.

Revealing Information

The director's job is to use shots and sequences to reveal information at the right time and in the right way to tell the story. Sometimes the audience is given information at the same time as the characters in the story. Sometimes the characters know more than the audience, and at other times the audience knows more than the characters. How and when this information is revealed is determined by the script and by how the director chooses to tell the story.

Let's look at an example. You have a robbery at the beginning of the story. The director may choose to let the audience know who the robber is at the beginning, but the main character (the hero) has to figure it out over the course of the movie. Watching to see if the hero will catch the robber, or if the robber will get away, creates tension and drama. The audience cheers for the hero, but is frightened for her at the same time because we know that her boyfriend is the robber and he's going to get away with the robbery if she does not figure out who did it.

Another option is to let the audience find out who the robber is at the same time as the hero. This creates tension by keeping the audience guessing and being surprised with what the hero finds out. The story is a "Who done it?" and we try to solve the mystery before the hero. Either approach works, but the director needs to decide which is the best way to tell the story.

Working with Cast and Crew

Every actor is different, and that makes it impossible to tell you how to work with them. I have worked with a few directors and seen a few at work, and I have directed as well. The directors that people enjoy working with do two things: First, they respect the people they work with, not just the actors, but the crew as well. They give the actors all of the information they need in order to succeed, and then allow the actors to act. If the actor does not do quite what the director wants, the director helps them figure out how to improve without saying "do it my way." The directors I have enjoyed working with establish the fact that they are in charge, but talk instead of yell, and communicate clearly.

Second, the director knows what is going on and has a clear vision of what needs to happen. One set I worked on was frustrating because the director had no idea what he was looking for. He would have the camera crew setup then he would look at the shot. He would walk around and around the set, think for a minute, then have them move the camera. He would then decide he didn't like the scene and quickly re-write the scene. Each shot took a couple of hours with very little work getting done. Everybody, from the actors to the crew, became frustrated very quickly. The project turned out okay, but it took forever, went over budget, and nobody wanted to work with the director again. Make sure you have a vision, but be flexible enough to make changes when needed.

SUMMARY

In this lesson you learned:

- A director's most important job is to tell the story.

- A shot is the time between when you start and stop recording. It is the basic building block for a video.

- A sequence is a series of shots that are edited together to create a scene.

- A basic sequence consists of an establishing shot that establishes where objects and characters are in a scene; followed by a close-up of characters, while they are speaking; reaction shots to show how characters react to what is happening in the scene; and, cutaways to show other events in the scene.

■ Each shot that requires a new setup should show a clear change in both size and camera angle. The setup for each close-up following the establishing shot should be closer and show a noticeable change in angle.

■ Clean in and out makes editing on actions smoother. Showing a car leave the frame, then cutting to a shot of the destination and the car entering the frame helps editing the shots together to look more natural.

■ "The line" is an imaginary line that determines where the camera should be setup when shooting a sequence.

■ A simple sequence with a master shot and two close-ups requires three setups: one for the master shot, and two for each of the close-ups.

■ The cameras for a multi-camera shoot should be setup so that the shot from each camera should show change of both image size and angle.

■ A multi-camera shoot takes several sources and puts them through a switcher. The switcher then allows the director to determine which of the sources will be shown or recorded at any given time.

■ Tension and drama partially are created by how the director builds sequences to reveal information to both the characters in the story and the audience.

■ A director should have a vision for how the video will be made and then respect the cast and crew and allow them to do their jobs.

VOCABULARY *Review*

Define the following terms:

Continuity	Master shot	Shot
Cutaway	Matching time code	Switcher
Establishing shot	Program	Tape operator
Graphics operator	Reaction shot	Technical director
"The line"	Sequence	VTR (video tape recorder)

REVIEW *Questions*

TRUE/FALSE

Circle T if the statement is true or F if the statement is false.

T F **1.** The director's most important responsibility is to tell the story.

T F **2.** A shot is the time between when you start and stop recording.

T F 3. A viewer can look at a single shot for up to 30 seconds.

T F 4. The master shot is the same as a close-up.

T F 5. The master shot is the same as an establishing shot.

T F 6. Shooting to edit means the editor is on set at all times.

T F 7. A cutaway is the same as a reaction shot.

T F 8. "The line" helps determine where the camera should be setup.

MULTIPLE CHOICE

Circle the correct answer.

1. A sequence is
   A. a series of shots edited together
   B. a series of scenes
   C. the basic building block for video
   D. a number of pages in a script

2. A master shot
   A. must always be the first shot in a scene
   B. is the last shot in a scene
   C. could be the only shot in the scene
   D. is always the longest shot in a sequence

3. The basic sequence for a two person conversation
   A. requires at least 5 camera setups
   B. requires at least 7 setups
   C. requires a setup for each line of dialogue
   D. requires just 3 setups

4. The basic sequence could be described as
   A. master shot, close-up, master shot
   B. master shot, reaction shot, close-up
   C. master shot, close-up, close-up
   D. close-up, master shot, close-up

5. The basic setup for a multi-camera shoot
   A. is nothing like setting up a basic sequence
   B. uses 3 cameras
   C. requires only 2 cameras
   D. give two separate master shots

6. Crossing "the line" can be accomplished by
   A. cutting from the master shot to a close-up
   B. changing the position of the image
   C. a continuous camera move
   D. simply changing image size

7. Continuity
   A. is overrated and unimportant
   B. is unimportant for small shoots
   C. means each sequence is shot in full from every possible camera angle
   D. can be accomplished by changing image size and angle

WRITTEN QUESTIONS

1. Explain "the line" and how it helps the audience.

2. What does clean in and out make easier and give an example of how it can be used.

3. Explain changing image size and angle.

4. Explain how a multi-camera shoot is setup.

PROJECTS

PROJECT 8-1

Rent a movie and watch it. Pay attention to how each scene is setup. Does each scene follows the master shot, close-up formula, or does it vary? How does changing the sequence formula alter the emotion or feeling of the scene or sequence?

PROJECT 8-2

Shoot a sequence. It does not have to be long, but shoot something using the change size and angle rule. Then shoot the same sequence and ignore the rule. Look at the footage, and, if you can, edit the shots together. How does following the sequence formula change how the shots look and feel?

PROJECT 8-3

The sequence for a conversation is pretty easy, but what about a girl playing basketball? How would you build a sequence for something like that? Write down some ideas of what shots you might want to use to build that sequence. Go ahead and create a scene of a girl shooting baskets by herself. Determine who she is, why she's there, and when (time, date, time of the year, etc.) she's there to help you create this scene.

PROJECT 8-4

Watch a sporting event or news broadcast. Does the broadcast follow the master shot formula? How many cameras are being used? Where are the cameras placed? How smoothly does each shot fit together? Do the shots work together?

 ### WEB PROJECT

Find a video on the Web (YouTube, Google Video, for example) that an amateur has made. The video preferably should have a story to it. Pay attention to whether or not the director used the ideas discussed in this lesson. Determine how building a sequence helped, or would help, the video.

 ### TEAMWORK PROJECT

Team up with at least two class members and each shoot a scene, using the other two as actors (if you don't have other class members, use two different objects, such as a football and baseball). Video tape a conversation between the two actors or objects (record the audio for two objects as if they were talking together). Look at the footage, or (if you can) edit it together. Is there enough of an angle change between the master shot and the close-up? Shoot the conversation a couple of different ways, using different angles, and determine which angles provide changes in size and which do not. Also note the changes in the effects that different angles produce.

CRITICAL *Thinking*

ACTIVITY 8-1

Watch a feature film, preferably a classic like Gone with the Wind, Star Wars, something well known and successful. Pay attention to the way each scene is structured. Note how many times the scene follows the master shot sequence covered in this lesson, and how many times it does not. What is the effect of changing from the master shot sequence? Does it reveal information differently? Does it hurt or help the sequence? Try to get into the head of the director and try to figure out why he or she chose a different way to structure the scene. Was it a good or bad idea?

POSTPRODUCTION

Unit 3

Estimated Time for Unit: 5 hours

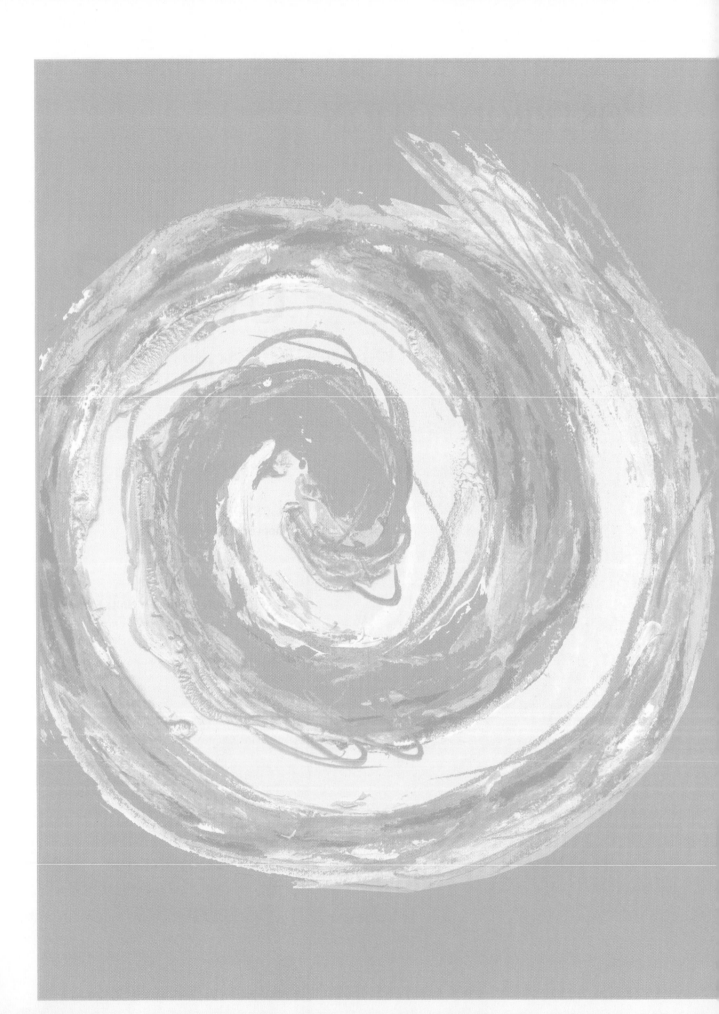

Editing Video

OBJECTIVES

Upon completion of this lesson, you should be able to:

- Connect your playback deck or camera to the computer
- Start Final Cut Express
- Create settings for an editing project
- Open a new project file
- Save your project file
- Capture your video
- Import video
- Create a new timeline
- Put video into the timeline
- Perform a basic cut
- Add transitions
- Render clips and transitions
- Add graphics and titles

Estimated Time: 1.5 hours

VOCABULARY

Capture

Editing

FireWire

Frame accurate editing

HDV

In point

Out point

Real time

Rendering

Time code

Trim or Trimming

USB

Once you have all of your footage shot it is time to put it all together to make a finished video. You will not want to use every second of video you shoot, so you need to edit it. Editing is the process of combining footage that tells your story and eliminating the footage you do not want. If you followed all of the guidelines and "rules" your shoot should have gone smoothly, and your edit should go smoothly as well. Even if you did do everything you learned about it is more than likely that you ran into some problems, maybe even some pretty big ones. That is okay, for some reason that is how it goes when you are working with video.

Connect Your Playback Deck or Camera to the Computer

DV (digital video) is a digital format, but the footage needs to be transferred to the computer so you can work with it. Most cameras use videotape to record and store footage, while other cameras use a hard drive instead. Transferring video from a hard drive allows users to simply copy the footage from a hard drive to the computer, or even edit the video right on the camera hard drive.

Videotape requires you to capture footage, which is basically recording the footage to a hard drive. The first thing you will need to do is connect your deck, camera, or playback device to the computer. Make sure you have everything you will need, which is a playback device with a DV (or FireWire) port, a FireWire cable, and a computer with a FireWire port. Remember, FireWire has a few other names, like IEEE 1394, DV, and Sony iLink, but no matter the name, it is an interface standard for connecting devices, like cameras and hard drives, to a computer. Most DV cameras and decks have a port that accepts 4-pin FireWire connectors, while many (but not all) computers have a port that accepts 6-pin FireWire connectors.

One of the great joys you will find in working with video is what I like to call the format wars. It is really fun to try and figure out if this deck can be connected to that video capture card, or if this video format will work with that video format. All DV cameras have

> **Note**
>
> Some cameras use USB connectors instead of FireWire. **USB** is an interface standard, like FireWire, but the two aren't interchangeable. It does not matter which type of connector you use, as long as you can connect the playback device to the computer. Just make sure you have the right computer equipment for the camera you use. Remember, FireWire has two types, 400 and 800, and USB comes in version 1.0 and version 2.0. FireWire 800 transfers data twice as fast as FireWire 400, while USB 2.0 is faster than USB 1.0.

FireWire ports but not all computers have them. All Apple computers have FireWire ports, but not all PCs do. Some PCs have FireWire ports, but many use 4-pin, not 6-pin. All computers have USB ports, but only a few cameras have them. In short, make sure you can connect your computer to your camera or deck before you buy anything.

Also, make sure you have the right software for the computer system you are working on. Final Cut Express (FCE) is a stripped down, lower price version of Final Cut Pro. Like Final Cut Pro it is only available for Apple computers. Adobe Premier Elements, however, is cross platform, meaning it will work on either Apple or Windows machines. Luckily the concept for working with both software packages is the same: capture the video, edit the video, and then put it out to a tape, DVD, or Web format (which we'll cover in Lesson 11). The first section covers Final Cut Express HD for the MAC and the second section covers Adobe Premier Elements for Windows XP.

Creating Settings for an Editing Project and Opening a New Project

Now that you have connected your computer to your deck or camera, you are ready to set up and use the software. The set up is important because it tells the software what format the video will be in when it is captured to the computer. For example, if the software is set up to edit DV footage, but you capture HD footage, you will have to render it and make it into DV footage in order to work with it. Final Cut Express works specifically with HDV, which is an MPEG 2 compressed high definition format.

The last thing you will need to do to set up the computer is to tell the computer where you want to put the footage you are going to capture. The location is known as a scratch disk, which is the common term for all video editing software. The default setting for Final Cut Express is HD (or whatever the hard drive is named) > Users > Default User > Documents > Final

> **Note**
>
> This book is designed to talk about DV, not HDV or other formats you might find. You can still follow this step by step if you shot the video with an HDV camera. Instead of setting up for DV footage, simply select the HDV format, such as HDV 1080i60 or HDV 720 30p. If you are not sure what these all mean, review Lesson 1.

Cut Express Documents. This is good for most cases, but you may find you want to change the scratch disk so you can capture to a different hard drive or an external drive.

STEP-BY-STEP 9.1

In this Step-by-Step activity, you create the settings for your editing project. You select the DV set up options so that the DV video you shot will work like it should while you are editing. If you do not set the project up correctly, you have to render every thing you are working with, which takes up a lot of time. You also tell the computer where to put the footage you capture.

1. Turn on your computer and camera.

2. Double-click the **Final Cut Express HD** icon to launch the software, as seen in Figure 9-1. If this is the first time you have started Final Cut Express you will see the Easy Setup Dialog box, as seen in Figure 9-2. This dialog box will tell the software what tape format you will be working with and how it will communicate with the playback deck.

FIGURE 9-1
Launching Final Cut Express

STEP-BY-STEP 9.1 Continued

3. The Dialog box should show DV-NTSC, as seen in Figure 9-2. If any other option is showing, click the arrows on the right side of the window to show a drop-down menu. Select **DV-NTSC** from the menu. If you do not remember what NTSC is and the reason you want to select that option you should review Lesson 1.

FIGURE 9-2
Easy Setup Dialog box

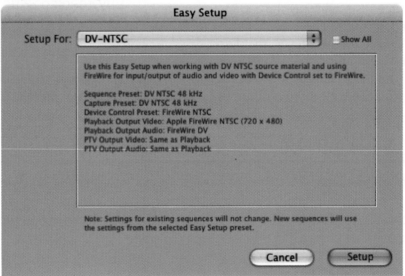

4. Click **Setup**.

5. If you launched Final Cut Express earlier and did not select the right format or do not remember what format you selected, you can open the Easy Setup dialog box from the menubar by selecting **Final Cut Express**, selecting **Easy Setup**, and then following steps 3 and 4.

6. Click **Final Cut Express HD**, then click **System Settings** and verify that the Scratch Disks tab is selected. Click the first **Set** button, as seen in Figure 9-3. This will bring up the Choose a Folder dialog box.

Note ✓

You may have noticed the easy set up choice DV-NTSC FireWire Basic. Earlier we talked about format wars and that not every type of tape deck or camera is compatible with every computer. Just because a deck or camera has a FireWire port does not mean it will work well with every computer with a FireWire port. If you connect your camera to the computer with FireWire and it does not seem to work right, switch the Easy Setup dialog box to DV-NTSC FireWire basic.

STEP-BY-STEP 9.1 Continued

FIGURE 9-3
Scratch Disks Set button

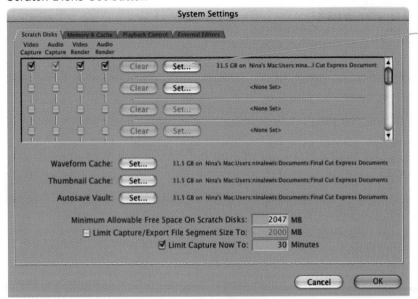

**Scratch Disks
Set button**

7. You can select the hard drive you want from the frame on the far left or you also can create a new location on the same hard drive. This might come in handy for each new project you create in order to keep track of where your footage is.

8. Next you create a new folder. Check with your instructor regarding where it should be saved, navigate to that location, then click **New Folder**.

9. When the New Folder dialog box appears, type **Digital Video Solutions**, then click **Create**.

10. Click **Choose**. This automatically will create three folders inside of the Digital Video Solutions folder: Audio Render Files, Capture Scratch, and Render Files. All of the files you capture will be stored in the Capture Scratch folder.

11. Click **OK** to complete the process. Leave the program open for the next Step-by-Step activity.

Working with Final Cut Express Windows

Now that Final Cut Express is running let's look at the five basic windows you need to know in order to work.

The first window on the left is the Browser window, as seen in Figure 9-4. In the Browser window you will see two file tabs: one labeled Untitled Project 1 and a second labeled Effects. The Untitled Project 1 tab is the project tab. This tab is where any captured footage will be, as well as the timeline you will be editing in. (You can create any number of timelines for a project, which can be helpful when working on problematic sequences without affecting other parts of the final program). The Effects tab is where all of the transitions are kept.

FIGURE 9-4
Browser window

The center window on the top is the Viewer window, as seen in Figure 9-5. This is where you work on individual clips before you add them to your timeline. You can set in and out points for a clip, make audio adjustments, add motion to a clip, and make other adjustments through the Filters tab. You can also add different special effects to a clip, like adding motion, cropping, resizing, or any number of different effects.

FIGURE 9-5
Viewer window

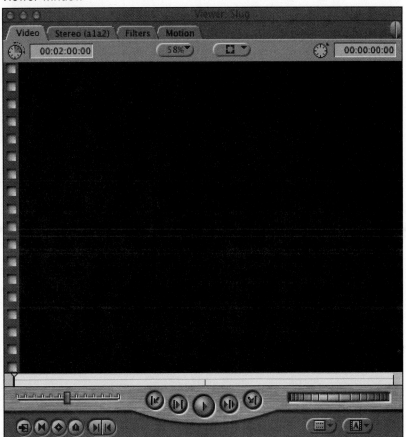

The window on the right is the Canvas, as seen in Figure 9-6. This shows you what is in the timeline so you can see what is edited together. It is important to remember the difference between the Canvas and the Viewer windows and what you are looking at.

FIGURE 9-6
Canvas window

The big window on the bottom is the timeline, as seen in Figure 9-7. This is where the clips you add to your edit are put together. The top two tracks are the video tracks, V1 and V2, while the bottom four tracks are audio tracks, A1 through A4. You can have up to 99 audio tracks and 99 video tracks in the timeline at the same time, but if you do, your computer is going to have a hard time playing all of those tracks.

FIGURE 9-7
Timeline

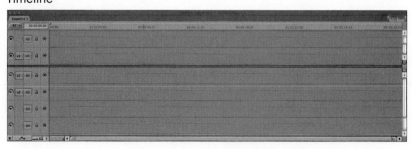

Save Your Project File

Before we get too far, it is a good idea to name and save the project you are working on. It is always a good idea to give each project a unique name so you do not have to guess which project is which.

STEP-BY-STEP 9.2

In this Step-by-Step activity, you save your project file.

1. Click **File**, then click **Save Project As**.

2. Click the drop-down box next to the **Where: field** and click **Digital Video Solutions** folder.

3. In the Save As: field, type **SBS9-2**, then click **Save**.

4. Leave the program and file open for the next Step-by-Step activity.

Capturing Your Video

You are now ready to capture video. There are three different ways you can capture video in Final Cut Express: capture clip, capture now, and capture project. Capture clip and capture project allow you do identify and capture only the parts of a tape you want to use, while capture now captures everything on the tape, whether you want it or not. There are times when the only way you will be able to capture footage is by using capture now. In this chapter, you use two of the ways to capture video: capture clip and capture now.

For this section it is important to remember what time code is. Time code was covered earlier, but a quick review will not hurt. Each frame of video is given its own unique name or identifier. The name is shown in hours, minutes, seconds, and frames. For example, the time code 03:34:18:19 is 3 hours, 34 minutes, 18 seconds, and 19 frames. If you want to have video from 03:32:12:18 to 03:34:18:19 in your video, Final Cut Express would cut out everything after 03:34:18:19. Without time code, it is impossible to create frame accurate edits.

STEP-BY-STEP 9.3

If you have video to capture, then in this Step-by-Step activity, you capture video using capture clip. First you identify the clip you want, and then capture the clip itself. If you do not have video to capture, then go to Step-by-Step 9.6 Import Video.

STEP-BY-STEP 9.3 Continued

1. In the file named SBS9-2, open the Capture window by clicking File, then Capture, as seen in Figure 9-8. The window on the left is the Preview window that displays what you are capturing. If the deck is connected to the computer (and the deck is on) you will see the transport controls that allow you to control the deck. The window on the right is the Capture file tab that allows you to log clips and identify information and notes about each clip.

FIGURE 9-8
Opening the capture window

2. Click the **Play** transport control, as seen in Figure 9-9. If the deck is connected correctly (and you put a tape with footage on it into the deck) you will see the video footage in the Preview window.

FIGURE 9-9
Play on the Preview window

STEP-BY-STEP 9.3 Continued

3. Select the time code for the in point of the video by clicking the Mark In button below the Preview window. The time code you selected will appear in the in point window, as seen in Figure 9-9. This means the computer will start to capture the video (and audio) at that time code.

4. Let the tape continue playing for a few seconds and click the Mark Out button, as seen in Figure 9-9. The out point, or where you want to stop the video capture appears in the out point window.

5. Click Clip in the Capture area on the bottom of the Capture window, as seen in Figure 9-10.

Note

Some decks will give you deck control through the computer as soon as the two are connected; others require you to select how you will control the deck. This usually is done by flipping a switch on the front of the deck. Look for a switch or a button on the front of the deck that is labeled something like local (to use the controls on the deck) and remote (to use the controls on the computer). Sometimes the controls on the deck won't work if remote is selected. If all else fails, read the user's manual for the camera or deck you are using.

FIGURE 9-10
Capture clip button

Capture Now button

Capture Clip button

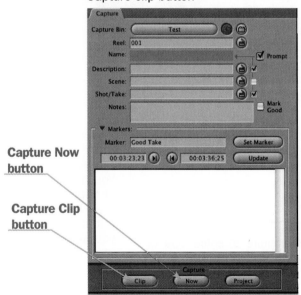

STEP-BY-STEP 9.3 Continued

6. A Log Clip window will appear and prompt you to name the clip as seen in Figure 9-11.

FIGURE 9-11
Log Clip window

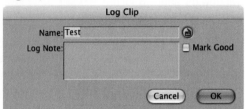

7. Type **Test** and click **OK**. The capturing window appears on the screen and the deck (or camera) will rewind the tape to the beginning time code of the clip and start the capture.

8. When the end time code is reached the capture will stop. A clip named test will appear in the Browser window, as seen in Figure 9-12.

FIGURE 9-12
Test clip in the Browser window

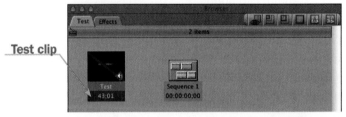

9. Click **File, Save Project As** and name the **file SBS9-3** and if possible, leave the file open for the next activity.

STEP-BY-STEP 9.4

Sometimes it is easier to just capture an entire tape, or watch the footage and capture as you go along. The easiest way to do that is to use capture now. Other times you will have time code issues, meaning Final Cut Express cannot identify the time code you are looking for, or does not have enough time code before or after the clip to work with. Capture Now captures the footage without concern for the time code.

1. With the file named **SBS9-3** open, turn on your video camera or deck if it is off.

2. To open the Capture window if it is not open already, click **File**, then **Capture** from the menu bar.

> **Note**
>
> You need at least 5 seconds of continuous time code before and after the clip you logged which is called pre-roll and post-roll. If you shoot for a minute, then stop for a minute, then start shooting again you will have a time code break: The time code starts then stops, then starts again. If you do not have enough pre-roll and post-roll Final Cut Express will get confused when it hits the time code break and won't be able to locate the clip. Do not worry, if you do run into this problem you will still be able to capture what you need.

STEP-BY-STEP 9.4 Continued

3. To play the tape, you can simply press the **Play** button on the deck, or you can click the **Play** button in the transport controls.

4. When the tape starts playing click **Now** at the bottom of the capture section as seen in Figure 9-10. The Capture window will come up and the capture will begin.

5. Press **ESC** on the keyboard to stop the capture. A clip named **Untitled** will appear in the Browser window, as seen in Figure 9-13. In order to make things easier to understand, later we will refer to the captured footage in the Browser window as a captured clip.

> **Note** ☑️
>
> Final Cut Express automatically names the clip Untitled when you use capture now. A second clip would be named Untitled1, a third clip would be named Untitled2, and so forth. You can rename the clips by clicking on the file, waiting a second, and clicking again. A box will appear around the name and the name will be highlighted in blue. You can type in the new name and it will replace the old name. If you click too fast (double-click) you will open the clip.

FIGURE 9-13
Untitled clip in Browser window

Untitled clip →

6. Save the project as **SBS9-4** and if possible, leave the project open for the next activity.

Import Video

What if you have a video that you captured for a different project, or you have a graphic that you made in Photoshop that you want to add to your project? Any number of media assets can be added to the project by importing the file into Final Cut Express. Importing video is also the way to get video shot on a camera into Final Cut Express.

STEP-BY-STEP 9.5

1. With the file named **SBS9-2** opened, click **File**, then **Import**, and then **Files** from the menu bar. This brings up the **Choose a File** window. You will see the list of hard drives, folders, and locations.

STEP-BY-STEP 9.5 Continued

2. Click the Digital Video Solutions folder , then click the file named **Colonial Parade.Avi**. Click **Choose**. The file is imported into the Browser window as seen in Figure 9-14.

3. Repeats steps 1 and 2 and import the file named **Italian Parade.avi**. Colonial Parade.avi and Italian Parade.avi clips appear in the Browser window as seen in Figure 9-14.

FIGURE 9-14
Importing files into the Browser window

Imported clips appear in the Browser window

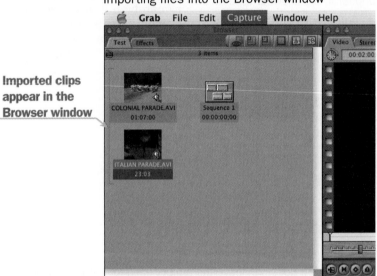

4. Save the project file as **SBS9-5** and leave it open for the next Step-by-Step activity.

Trimming Clips

Before you add any clips into the timeline you need to trim away any parts of the captured clip that you do not want included in the final edit. This means you need to set an in point and an out point for each captured clip in order to create an edited clip.

There are two ways to load a captured clip into the Viewer window: double-click the captured clip, or click and drag the captured clip from the Browser window.

S TEP-BY-STEP 9.6

1. With the file named **SBS9-5** open double-click the captured clip named **Colonial Parade.avi** located in the Browser window or you can click and drag the clip to the Viewer window.

2. Press the **Play** button on the bottom of the Viewer window (once you press the play button, it turns the button into a yellow arrow) or press the spacebar on the keyboard to play the clip. You can also quickly scrub through the clip by dragging the playhead through the scrubber bar, as seen in Figure 9-15.

FIGURE 9-15
Dragging the playhead through the scrubber

3. Press the **Stop** button at the bottom of the Viewer window (this is a yellow arrow) or the spacebar on the keyboard when the video is close to the timecode **00:00:07:00**. This is where you want the clip to start. Use the right and left arrows on the keyboard to move frame by frame until you get to timecode **00:00:07:00**.

Did you know?

The timecode reader is located at the tip right of the viewer window.

Did you know?

You can also press the **I** key on the keyboard, or click **Mark**, then **Mark in** from the menubar to mark the frame as the first frame you want to use.

STEP-BY-STEP 9.6 Continued

4. Click the **Mark In** button at the bottom of the Viewer window as seen in Figure 9-16.

FIGURE 9-16
Mark In button at the bottom of the
Viewer window

**Timecode
reader**

**Mark In
button**

> **Note** ☑
>
> Notice the in Point marker in the scrubber bar, and that the area behind the in Point in the scrubber bar is darkened, as seen in Figure 9-17.

FIGURE 9-17
Mark In marker on the scrubber bar

**Mark In marker on
the scrubber bar**

STEP-BY-STEP 9.6 Continued

5. Click the **Play** button, or scrub through the clip by dragging the playhead through the scrubber bar until you reach time code **00:00:45:00**. You can also use the right and left arrows on the keyboard to move frame by frame until you reach time-code **00:00:45:00**.

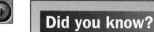

> **Did you know?**
>
> You can also press the **O** key on the keyboard, or select **Mark**, then **Mark Out** from the menubar to mark the frame as the last frame you want to use.

6. Click the **Mark Out** button at the bottom of the Viewer window. Notice the area in front of the out point in the scrubber bar is darkened, as seen in Figure 9-18.

FIGURE 9-18
Mark Out marker on the scrubber bar

Mark Out marker
on the scrubber bar

7. Save the project as **SBS9-6** and if possible, leave the project open for the next Step-by-Step activity.

Putting Video into the Timeline

Now that you have marked in and out points in a clip, it is time to add the clip to the timeline. There are a couple of different ways to add the clip to the timeline. You can simply drag the clip into the timeline, or pull it over to the Canvas window. If you pull the clip over to the Canvas window, a few options will slide in from the right hand side of the window. Each option places the video in the timeline in a different way, as seen in Figure 9-19.

Note

It is important to remember that setting in and out points does not affect the original captured clip. The software keeps the entire clip, but only plays the clip from the in point to the out point once it is in the timeline. You can go back to the same captured clip, set new in and out points, and use it later in the timeline without affecting the first time you used the clip. It is like reading a few pages from a book, then putting the book down, reading a few pages from another book, and then going back and reading different pages from the first book. The unread pages aren't destroyed, and you can go back to that book again and again.

FIGURE 9-19
Canvas window menu items

Insert with transition adds the default transition (which is a dissolve). Fit to fill alters the time of the clip so, for example, a 30 second clip will be sped up to play in just 20 seconds, or slowed down to play in 45 seconds. The function of each of these options can become quite confusing, so make sure you understand what each function will do before you try it. Remember, though, that you can use Command+Z or select Edit, then Undo from the menu bar any time you do something you did not want to do.

STEP-BY-STEP 9.7

In this next Step-by-Step activity, you try both of the methods for adding a clip to the timeline.

1. With the file named **SBS9-6** opened, click and drag the Colonial Parade clip from the Viewer window down to the Timeline next to the v1 track. This places the video clip beginning at the position of the playback head in the timeline.

2. To try the second method of adding a clipto the timeline, we need to import another clip. Repeats steps 1 through 6 from Step-by-Step 9.7 for the Italian Parade clip. Set a Mark in point at timecode **00:00:06:00** and set a Mark out point at timecode **00:00:14:00**.

3. Click and drag the clip from the Viewer window to the Canvas window. Drag the clip over Insert and release the mouse button. The clip will be added to the timeline as seen in Figure 9-20.

> **Important**
>
> It is important to remember that the position of the playback head determines where the clip will be placed in the timeline. If the playhead is in the middle of a clip already in the timeline, the new clip will overwrite the second half of that clip.

> **Important**
>
> When you drag the clip to the Viewer window a menu will slide into the window from the right. The options are insert, insert with transition, overwrite, overwrite with transition, replace, fit to fill, and super impose.

FIGURE 9-20
Click and drag the clip from the Viewer window to the Canvas window

Click and drag the clip from the Viewer window to the Canvas window

4. Save the project as **SBS9-7** and if possible, leave the project open for the next Step-by-Step activity.

Adding a Transition

Most edits you see in movies and on television are simply straight cuts, meaning the last frame from the first clip is followed directly by the first frame of the second clip. Most conversations, for example, are edited together using simple straight cuts. Every so often however, a straight cut just does not do it; imagine a romantic scene from a movie full of hard cuts.

Transitions are just what they sound like, a transition between two shots. Transitions can smooth out differences between shots, convey the passage of time, or a change in location. Pay attention to the transitions in a movie or television show and see how they are used.

Generally speaking, you want to edit in a way that does not call the audience's attention to the editing itself. For example, audiences are used to the convention of editing close ups together to create the idea that two people are talking to each other. Most edits do not scream for attention. A slide or wipe, for example, calls the audience's attention to the editing, and you can use that to convey an idea, such as "we are now going to another place."

Most of the professional editors I have worked with use straight cuts and dissolve, and a wipe or something more distracting once in a while, depending on what they are trying to accomplish.

A transition basically takes frames from the first clip (called the outgoing clip) and blends them together with frames from the second clip (called the incoming clip). The default duration for every transition in Final Cut Express is a second, which is 30 frames. (Remember, DV plays at 30 frames each second). This all means that the transition will start 15 frames, or a half of a second, *before the out point* of the out going clip and will end 15 frames, or half of a second, *after the out point* of the out going clip. The transition will also begin 15 frames *before the in point* of the incoming clip, and end 15 frames *after the in point* of the incoming clip.

For example, let's say I have a captured clip that is 120 frames long and then I create an edited clip with an out point which is frame 118 of the captured clip. I add the edited clip to the timeline and put it together with a second edited clip. When I play what I have put into the timeline, the first edited clip will play until frame 118 and cut directly to the first frame of the second edited clip. If I add a transition between the first clip and the second clip it won't work. Why? When the first clip in the timeline sees that it needs fifteen frames *after* the out point in order to make the transition work, it goes back to the captured clip and adds fifteen frames after the out point of the edited clip. This means that I would need to add frames 121 to 135 in our 120 frame clip. Since the last frame in the captured clip is frame 120, however, the edited clip cannot add any frames to complete the transition because the frames do not exist. You can remedy this in a couple of ways; first, you can set the out point at frame 105 of the captured clip, or, second, you can change the duration of the transition.

One more note about transitions, notice in Figure 9-21 that Box Slide and Push Slide are in bold type, but that none of the others are in bold. The effects with bold face type are real time effects, and the effects in normal type face are not. **Real time** means that the effect is active as soon as you add the transition to the timeline. None of the real time effects must be rendered before they can be played. See the section on Rendering video clips and transition in this lesson.

FIGURE 9-21
Slide folder under Video Transition on
the Effects tab

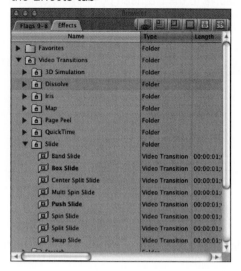

STEP-BY-STEP 9.8

In this Step-by-Step you place the playhead between two clips, and then add a simple transition between them.

1. With the file named **SBS9-7**, make sure the playhead is positioned at the end of the first clip and before the second clip as seen in Figure 9-22.

FIGURE 9-22
Positioning the playhead for the transition

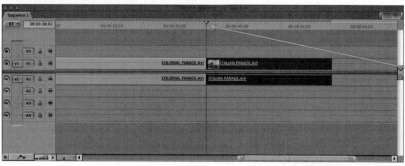

Positioning the playhead for the transition

2. Click the **Effects** tab in the Browser window.

3. Click the triangle next to the **Video Transitions** folder.

4. Click the triangle next to the **Dissolve** folder.

> **Note**
>
> You also can position the time-line playhead by clicking the **Go to Previous Edit** or **Go to Next Edit** buttons at the bottom of the Canvas window or you can use a key short cut. The ; key will move the playhead to the previous edit, while the ' key will take you to the next edit.

STEP-BY-STEP 9.8 Continued

5. Click and drag **Cross Dissolve** to the playhead point in the timeline as seen in Figure 9-23.

FIGURE 9-23
Adding a Cross Dissolve transition to the timeline

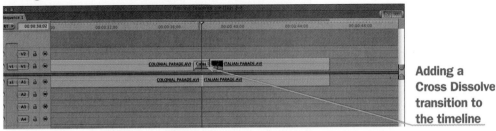

Adding a
Cross Dissolve
transition to
the timeline

6. Click the SBS 9-7 tab in the Browser window.

7. Save the project as **SBS9-8** and if possible, leave the file opened for the next Step-by-Step activity.

Congratulations. You have added a transition to the timeline. Notice back in the Browser window that Cross Dissolve is underlined. This means that cross dissolve is the default transition. You can add the default transition to the time by moving the playhead to the edit point in the timeline where you want to add the transition and then pressing command T on the keyboard. The reason you learned the long way is so you can see the other transitions that are available and figure out how to add other transitions.

Rendering Video Clips and Transition

Y ou may notice on your timeline that there is a red line across the top of the timeline. This means that the clips and transition have not been rendered. Rendering means that the video is processed so that it can play correctly and smoothly. Rendering a transition, for example, takes frames from clip A and combines them with frames from clip B to create new video frames.

S TEP-BY-STEP 9.9

In this Step-by-Step activity, you will render the clips and transitions.

1. With the file named **SBS9-8** opened, click anywhere on the Timeline.

2. Click **Sequence**, point to **Render All**, then click **Both** from the menu bar or press **Alt-R** on the keyboard. This will take about a minute to render. The bar on the top of the timeline appears green or blue.

3. Save the project as **SBS9-9** and if possible, leave the file open for the next Step-by-Step activity.

> **Note**
>
> You can, however, render specific pieces of the video. For example, if you are editing along and want to see how the transition you just put in looks, then you will probably want to render only that transition.

Adding Graphics and Titles

You can now do everything you need to do to edit in Final Cut Express: capture video, import captured footage, set in and out points, place footage in the timeline, and add transitions. Adding titles and graphics makes your project look nicer and more professional.

STEP-BY-STEP 9.10

In this next Step-by-Step activity, you add a simple text graphic to your timeline and play your video.

1. With the **SBS9-9** opened, click the **Effects** tab in the Browser window.

2. Click the arrow to the left of the folder labeled **Video Generators**.

3. Click the arrow to the left of the folder labeled **Text**.

4. Double-click the icon labeled **Text**, as seen in Figure 9-24. This will load a new title into the Viewer window.

FIGURE 9-24
Text icon on the Effects tab

Text icon on
the Effects tab

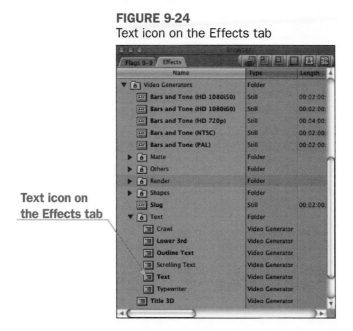

STEP-BY-STEP 9.10 Continued

5. Click the **Controls** tab at the top of the Viewer window as seen in Figure 9-25.

FIGURE 9-25
Controls tab on the Viewer window

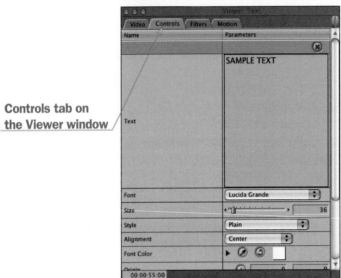

Controls tab on
the Viewer window

6. At the right of the Controls window is the Text window. Click in the text window to select **SAMPLE TEXT**.

7. Type **My Cool Title**. Below the text window are the font controls that allow you to change font, size, position, color, etc. Go ahead and change whatever you want, and if you do not like what happens, just use undo.

8. Click the **Video** tab in the Viewer window when you have finished with the text. The default text is white with a transparent or black background.

9. Click in the Viewer window and drag the frame to the timeline. You should have video in the timeline already. If you drag the text frame to the beginning of the video track one, labeled **v1**, the text appears over a black background as seen in Figure 9-26.

FIGURE 9-26
Adding the title to the beginning of the timeline

Adding a title
to the
beginning
of the
timeline

STEP-BY-STEP 9.10 Continued

10. To play your video, position the playhead at the beginning of the timeline by pressing the ; key on the keyboard.

11. Click the **Play** button on the Canvas window. Look for the Cross Dissolve transition as you play the video.

12. Save the project as **SBS9-10**. Close the project file and exit Final Cut Express.

Once the video is edited you can put the edited video on a tape, DVD, or the Web. We will cover how to do deliver your edited video to various outputs in Chapter 11. The next section of this lesson covers Adobe Premiere Elements for Windows XP.

> **Note**
>
> If you drag the text into video track two (v2) the text will be superimposed over any video shown at the same time in track one. Superimposed means that the text will be seen over the top of whatever video is shown at the same time. You also can put the video at the beginning of the track for the beginning titles, at the end for the end titles, or wherever.

Working with Adobe Premiere Elements

Adobe Premiere Elements is an easier program to set up and use than Final Cut Express, but it does not offer the same tools and control. Premiere Elements does offer tools for creating DVDs, while Final Cut Express requires the use of another software package. Final Cut Express can only be used on Macintosh computers, which can be a problem if you do not have one.

The process for editing in Premiere Elements is the same as it is for Final Cut Express. In fact, the process is the same for every editing system I am aware of. First, you capture the video, second you trim the clips, third you add them to the timeline, and finally you add transitions and titles.

Connecting Your Playback Deck or Camera to Your Computer

Connect the camera or playback deck to your computer the same way you did with the Macintosh. FireWire works the same on a PC as it does on the Mac, as well as USB. Most cameras have FireWire connections, although it is more uncommon (although becoming more common) to find USB connections, or ports. Make sure the computer and camera have the right connections.

Before you begin, make sure you have correctly connected your camera or video playback deck to the computer. If you have any problems, make sure you have the correct, and usually most recent, FireWire device drivers installed on your computer. The following activities are done on a Windows XP platform.

STEP-BY-STEP 9.11

In this Step-by-Step activity, you launch the Adobe Premiere Elements program.

1. Click **Start** in the taskbar.

2. Click **All Programs**.

3. Click the **Adobe Premiere Elements 2.0** icon. This launches the program and brings up the welcome screen. It may take a few seconds before the welcome screen appears depending on the computer you are using. Note: You may also have an Adobe Premiere Elements icon on the desktop if you chose that option during installation. You can simply double-click that icon to launch the program.

The welcome screen gives you several options for how to begin the program as seen in Figure 9-27, and you can look at each one as you like, but to start off we'll create a new project. We also will look at how to name a project and create a folder so you know the location of the media and elements of the project.

FIGURE 9-27
Welcome screen

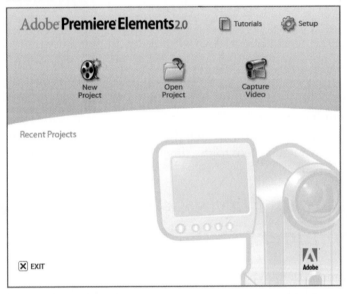

4. Click the **New Project** icon at the top left of the welcome screen. This brings up the New Project dialog box.

5. Type **My Project** in the Name box.

6. Click the **Browse** button on the right side of the New Project box. This brings up the Browse For Folder dialog box. This is where you set up where the project and media elements are stored on the computer.

7. Highlight the place where you want to project file saved. Check with your instructor.

STEP-BY-STEP 9.11 Continued

8. Click the **Make New Folder** button on the bottom left of the browse for folder dialog box. This creates a new folder inside of the highlighted folder you chose in Step 7. The name of the new folder is highlighted.

9. Type **Digital Video Solutions**, then click **OK**. This closes the Browse For Folder dialog box. Three folders are created under the Digital Video Solutions folder: Encoded Files, Media Cache, and Adobe Premiere Elements Preview files.

10. Click **OK**. This brings up the Adobe Premiere Elements work space. Leave the program open for the next Step-by-Step activity.

The Premiere Elements work spaces and panels are similar to what you find in Final Cut Express and most other video editing software. Elements can display several different panels, but this lesson will only cover four: the monitor panel, the media panel, the timeline panel, and the effects and transitions panel. Feel free to take a look at the other panels as you become more familiar with the software. One that can be very helpful is the How To panel. The panels can be organized however the user wants to display them, but we will discuss them according to the default positions.

The Monitor panel is positioned in the top center of the screen, as seen in Figure 9-28. The Monitor switches between displaying the video from the Timeline, or an individual clip. You can switch between the timeline and clip mode by clicking on the appropriate button at the top of the panel as seen in Figure 9-28. Make sure you know whether the Monitor is showing a clip or what is in the Timeline or you might change something you really did not want to change. For example, the in and out points for a clip can only be set when the monitor is in the clip mode.

FIGURE 9-28
Monitor panel

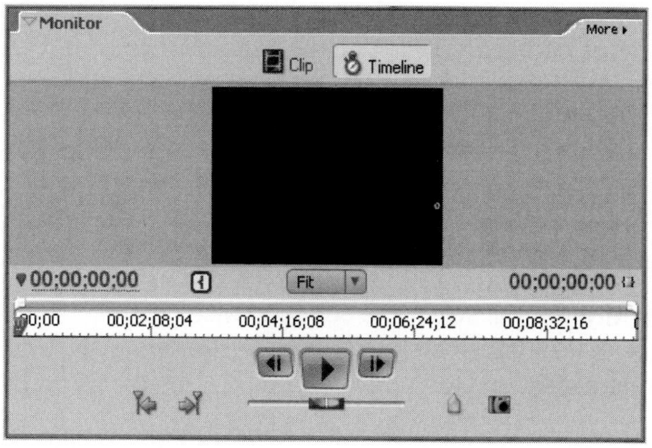

The Media panel is displayed on the upper, left-hand corner of the screen, as seen in Figure 9-29. This is where the media is placed for easy access. The media is displayed with an icon of the first frame of video (for video clips, that is) for easy identification.

The Effects and Transitions panel is positioned below the Media panel on the left-hand side of the screen, as seen in Figure 9-30. This panel displays the available transitions and effects. This lesson will cover transitions, but not effects.

FIGURE 9-29
Media Panel

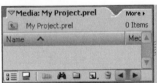

FIGURE 9-30
Effects and Transitions panel

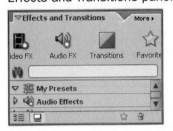

The final panel covered in this lesson is the Timeline, as seen in Figure 9-31. The Timeline is where the actual editing takes place. Trimmed clips are added to the timeline in the order you want them to be shown in the completed, edited video.

FIGURE 9-31
Timeline

Before you can capture video you need to make sure you have the correct device control selected for whatever type of camera you are using. Make sure to check the users manual if you are unsure of how the camera output works. The following activity shows you how to select the correct device control.

STEP-BY-STEP 9.12

In this Step-by-Step activity, you will select the correct device control.

1. Click **Edit**, then **Preferences**, and then **Device Control**. This brings up the Preferences dialog box.

2. Click the arrow next to the Devices drop down menu. This brings up the list of device control options.

3. Select **DV/HDV Device Control** if you are using IEEE1394, better known as FireWire, or **USB Video Class 1.0 – Device Control** if you are using USB device control.

4. Click **OK**. Leave the file open for the next Step-by-Step activity.

Now that you have set up the deck control you are ready to capture video.

STEP-BY-STEP 9.13

In this Step-by-Step activity, you are going to capture video.

1. Turn on your camera to the VCR or VTR mode. If you do not have video to capture, then go to the Step-by-Step 9.14

2. Click the **Capture** button just above the right-hand side of the Monitor panel as seen in Figure 9-32. This will load and display the Capture panel. Notice the message box at the top of the panel. This will let you know right away whether your deck or camera is connected to the computer.

FIGURE 9-32
Capture button

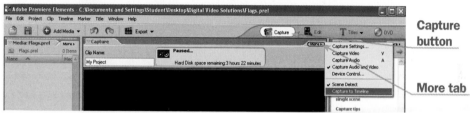

Capture button

More tab

3. Click the **More tab** on the right-hand side of the Capture panel, as seen Figure 9-32.

4. Click **Capture to Timeline** to deselect that option. The check mark next to capture timeline will disappear and the drop-down menu will close.

Note

If the message is something along the lines of Capture Device Offline, check your FireWire or USB connections and make sure the playback device is setup correctly. For example, most cameras have a playback mode setting in order for the device to playback what is on the tape.

Note

Premiere Elements places all captured clips directly into the timeline in the order that they are captured by default. Remember that most films are shot out of order.

STEP-BY-STEP 9.13 Continued

5. Highlight the text in the **Clip Name** field at the top of the Capture panel, as seen in Figure 9-33.

FIGURE 9-33
Changing the clip name

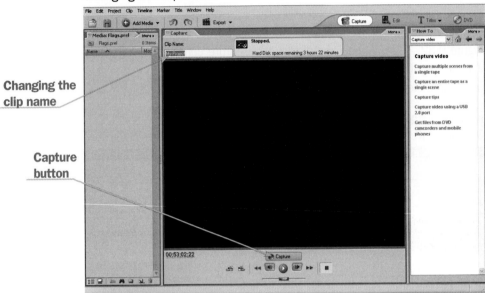

6. Click the green **Play** button at the bottom of the Capture panel as seen in Figure 9-33. The tape will begin playing. You can fast forward, rewind, and step through the footage using the buttons at the bottom of the panel.

7. Click **Capture** when you have located the footage you want to capture (See Figure 9-33). If you have stopped the tape, the Capture button will start the tape and capture the footage.

8. Click **Stop Capture**, as seen in Figure 9-34. This stops the capture process and places the captured clips in the media panel on the left of the screen.

> **Note** ☑️
>
> Make sure that when you capture you use a descriptive name, such as "Master shot A," or "Actor A close-up." That way you know what each clip is without having to watch the clip to see what is happening. It takes more time to name each clip this way, but in the end it saves much more time. Also, make sure that you capture only the footage you are sure you are going to use.

> **Note** ☑️
>
> Do not start capturing right on the first frame of video that you want. Make sure you have a few seconds of footage before the frame you want so you make sure you get the right beginning frame. Sometimes, the frame you want is right at the beginning of the tape and you do not have anything you can capture before that frame, so there's not much you can do about it.

STEP-BY-STEP 9.13 Continued

FIGURE 9-34
Stop Capture button

Stop Capture button

FIGURE 9-35
Renaming a clip

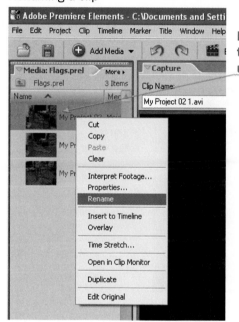

Right mouse click the clip to bring up menu list

> **Note** ☑
>
> If you did not name the clip before you captured it you can right-click the clip in the media panel and choose Rename from the menu that appears. Select Rename from the second section down as seen in Figure 9-35. That highlights the clip name and you can then type the appropriate name. You also can click the Clip name, then type the name of the clip.

9. Turn the camera off. Close the project file and do not save. Leave the project open for the next Step-by-Step activity.

Importing Media

There are times when you want to import media from another source instead of capturing it from a tape. For example, you may have a graphic or video clip you want to put in your video.

STEP-BY-STEP 9.14

In the next Step-by-Step activity, you import a video clip into your project.

1. To create a new project file, click **File**, then click **New, Project** and type **SBS9-14** and save it to the **Digital Video Solutions** folder. Click **OK**.

2. Click **File. Add Media**, then click **From Files or Folders**.

STEP-BY-STEP 9.14 Continued

3. Click the arrow next to the **Look in: text box**, then double-click **Digital Video Data Files** folder. Check with your instructor for the location of this folder.

4. Click the file named **Colonial Parade.avi**, then click **Open**. This adds the file to the media panel.

5. Repeat steps 1 through 4 except import the file named **Italian Parade.avi**. In your Media panel, you should see two clips as seen in Figure 9-36.

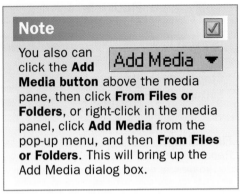

Note ☑

You also can click the **Add Media button** above the media pane, then click **From Files or Folders**, or right-click in the media panel, click **Add Media** from the pop-up menu, and then **From Files or Folders**. This will bring up the Add Media dialog box.

FIGURE 9-36
Importing video clips

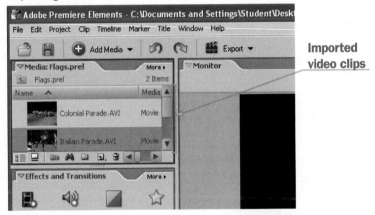

6. Save the project file as **SBS9-14**. If possible, leave the project file open for the next Step-by-Step activity.

Trimming Clips and Adding Clips to the Timeline

You are ready to begin editing once you have media to add to the timeline. The first step is to trim the separate clips and place them in the timeline. Trimming a clip removes unwanted frames from the clip. Trimming clips does not permanently remove frames from the captured video clip. The easiest way to think about it is that the software only shows the parts of the clip you want to show. Once the clip has been trimmed it is ready to add to the timeline.

STEP-BY-STEP 9.15

In this next step-by-step activity, you trim a clip to add to the timeline.

1. With the file named **SBS9-14** opened, click **Edit** next to the Capture button above the Capture panel. Notice that all of the captured clips are loaded into the media panel. If you chose to capture to the timeline the clips are placed in the timeline in the order you captured them.

2. To load a clip to the monitor, double-click the **Colonial Parade** clip or you can drag and drop it from the media panel to the monitor. Notice that the **Clip** button on the top of the monitor panel is highlighted, as seen in Figure 9-37. This means that all of the tools needed to trim the clip are active.

FIGURE 9-37
Loading a clip to the monitor

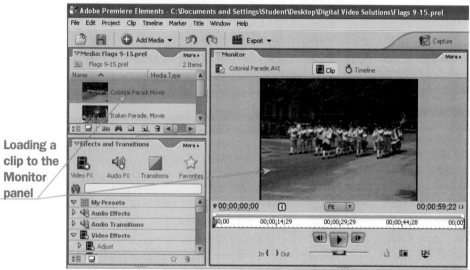

Loading a clip to the Monitor panel

3. Click the **Play** button at the bottom of the monitor or press the spacebar to move through the clip until you reach timecode **00:00:07:00**. Use the right or left arrow key on the keyboard to move through the clip frame by frame until you see the frame you want to use as the first frame.

4. Click the **Pause** button at the bottom of the monitor or press the spacebar.

STEP-BY-STEP 9.15 Continued

5. Click **Set In Point** button at the bottom of the monitor panel, as seen in Figure 9-38, or press the I key on the keyboard to set the current frame as the in point. This is now the first frame of that clip that the audience will see.

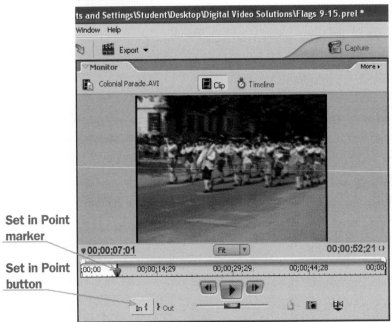

FIGURE 9-38
Set in Point

6. Play through the clip using the **Play** button or spacebar or use the arrow keys to reach timecode **00:00:45:00**.

7. Click the **Set Out Point** button at the bottom of the monitor panel, as seen in Figure 9-39, or press the O key on the keyboard to set the current frame as the out point. This is now the last frame of that clip that the audience will see. The clip has been trimmed and is now ready to be added to the timeline.

STEP-BY-STEP 9.15 Continued

FIGURE 9-39
Set Out Point

Set Out Point button

Set Out Point marker

8. Click and drag the clip from the monitor panel to the video1 timeline, as seen in Figure 9-40. This adds the clip to the timeline.

FIGURE 9-40
Dragging clip to Video 1 timeline

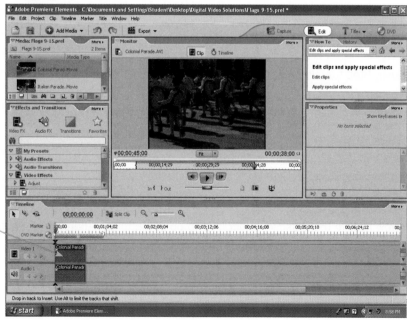

Dragging clip to Video 1 timeline

STEP-BY-STEP 9.15 Continued

9. Follow steps 2 through 8 and trim the **Italian Parade Movie** clip. Set in point at timecode **00:00:06:01** and set out point at timecode **00:00:14:00**. Add the trimmed clip to the timeline after the Colonial Parade clip as seen in Figure 9-41. The clips will snap to the previous clip when they are dragged to the timeline.

FIGURE 9-41
Adding a second clip to the timeline

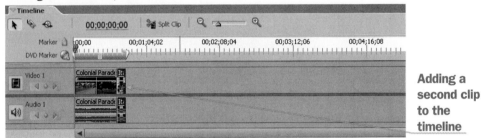

Adding a second clip to the timeline

10. Save the project as **SBS9-15** and if possible, leave it open for the next Step-by-Step exercise.

Adding a Transition and Rendering the Video

Once clips are added to the timeline, you can either leave the clips as they are with straight cuts, or add transitions between them. Sometimes it is easier to add transitions as you go along, while other times it is easier to put all of the clips in the timeline and add the transitions later. You do need at least two clips to transition between.

STEP-BY-STEP 9.16

In this next Step-by-Step exercise, you add a transition and render the video.

1. With the file named **SBS9-15** opened, make sure that you have two clips on the timeline next to each other and the playhead positioned between the two clips.

2. In the Effects and Transitions panel, click the arrow next to the Video Transitions folder icon. This opens the transitions folder and shows the transition types. You may need to use the scroll bar on the right side of the panel to see all of the transition types.

3. Click the arrow next to the Dissolve folder icon to open the folder. There are six different dissolves from which to choose. You may need to resize the Effects and Transition panel to see all six dissolves, as seen in Figure 9-42.

STEP-BY-STEP 9.16 Continued

FIGURE 9-42
Different dissolves templates

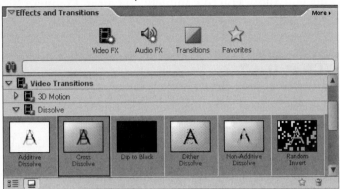

4. Click the **Cross Dissolve** icon and drag it to the seam between the two clips you want the transition. An arrow marks the seam between the two clips. Release the mouse button to add the transition to the timeline. You can use the arrow buttons or click and drag the playhead over the transition to preview the transition.

5. Make sure the Timeline button is selected at the top of the monitor window. If it isn't, click **Timeline**.

> **Note** ☑
>
> By this time you should have noticed that when you add a video clip to the timeline a red line appears at the top of the timeline. This means that the video needs to be rendered before the video will play smoothly. You can preview the video slowly, but it won't play smoothly until you render it.

6. On the menubar, click **Timeline**, then **Render Work Area** from the menubar, or press enter on the keyboard. The video will render out and the red line at the top of the timeline will turn green. Be patient, it may take a few minutes for the render to complete, depending on the length of the video, the number of video track you are using, and how complicated the transitions are.

7. After rendering is completed, the video plays in the Monitor panel showing the transition.

8. Save the project as **SBS9-16** and if possible, leave it open for the next Step-by-Step activity.

Adding a Graphic Title

So far you have captured video, trimmed clips, added clips to the timeline, and added transitions. These basic skills will allow you to complete an edit in Premiere Elements. The last thing we'll cover is creating titles.

STEP-BY-STEP 9.17

In the next Step-by-Step exercise, you create a title for your video.

1. With **SBS9-16** opened, click the **Titles** button at the top right-hand side of the program window. This opens the Templates window, as seen in Figure 9-43. On the left-hand side of the window are folders labeled with different types of title templates. Inside of these folders are variations of those themes. For example, Sports included three themes: Grind (BMX themed), Ski, and Soccer Action.

FIGURE 9-43
Templates window

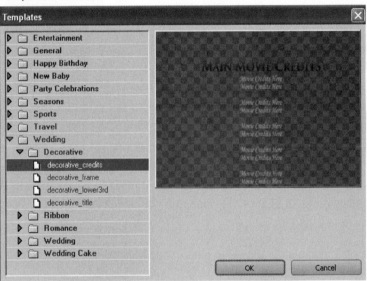

2. Click the arrow to the left of the folder labeled **General** to open the folder.

3. Click the arrow to the left of the folder labeled **Generic One**. This opens the folder and shows four templates based on the same theme, as seen in Figure 9-44.

STEP-BY-STEP 9.17 Continued

FIGURE 9-44
Generic One Title template

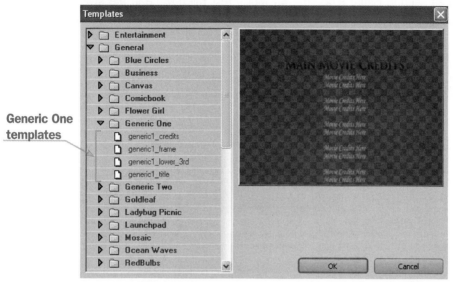

Generic One templates

4. Click the **generic 1_credits**. You can see what the template looks like in the frame on the right. The gray checkerboard shows the transparent area that video (or other graphics) will appear behind. Click **OK** to open the template in the Titler panel, as seen in Figure 9-45.

FIGURE 9-45
Generic 1_credits

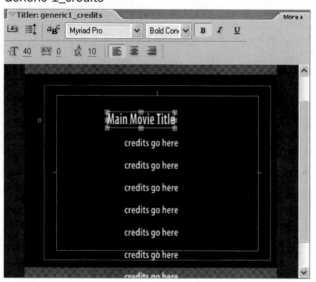

5. The **Main Movie Title** text box is selected. Highlight the text and type **My Movie**. The text "My Movie" is now the main movie title.

STEP-BY-STEP 9.17 Continued

6. Select the "My Movie" text and select a text style from the Titler Styles panel on the right hand side of the screen as seen in Figure 9-46. The My Movie text will now be in the style you chose. You can change the font, the font size, and other text attributes by using the menu at the top of the titler panel.

FIGURE 9-46
Text styles panel

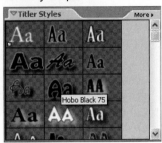

7. Select the next line of text and type your name. Select the next line of text after that and type your friend's name. Repeat for each line of text and adjust the text attributes in whatever way you want. Notice that the **generic 1_credits** has been added to the media panel.

8. Click the second button from the left in the Titler panel. This opens the Roll/Crawl Options window, as seen in Figure 9-47.

FIGURE 9-47
Roll/Crawl options window

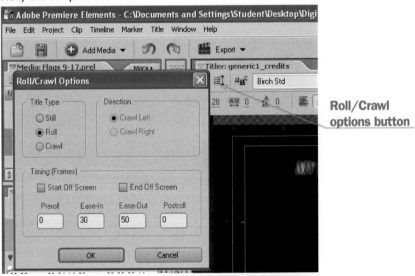

9. Click the circle next to **Still** in the **Title** Type box. This will make the title you created a still title as opposed to a rolling title. You can also create a crawl, which means the text will move from left to right, or right to left depending on which option you choose.

10. Click **OK**. The title is now saved as a still title and is loaded into your media panel. You can now add the title to the timeline by dragging the generic1_credit clip from the Media panel to the beginning of the timeline.

STEP-BY-STEP 9.17 Continued

11. Render the timeline by clicking **Timeline**, then **Render Work Area**, especially if the title has movement, such as a crawl or roll. You won't be able to see the movement unless it is rendered.

12. Play the video by clicking the **Play** button in the Monitor window.

13. Save the project as **SBS9-17** and exit Adobe Premiere Elements.

You can now edit a video using Adobe Premiere Elements. In the next lesson we will cover editing audio in Final Cut Express and Adobe Premiere Elements.

> **Note** ☑
>
> Earlier we talked about the transparency area behind (or under) the text. This means that you can add a title to the timeline and any video playing at the same time will appear behind the text. To try this out, add a video clip at the beginning of track one, and then add a title at the beginning of track two. Once you render the work area the video will appear at the same time as the title.

SUMMARY

In this lesson, you learned:

- How video must be captured, or transferred, onto the hard drive in order to work with it. The captured video must have the same settings as the sequence in order for the video to work properly without having to take time rendering.

- How to set up a project in Final Cut Pro and Adobe Premier Elements.

- How to set in and out points for the video using the keyboard and/or the mouse and how to trim out parts of the clip that are not needed for the edit.

- How transitions are added between clips to create new mixed frames of video. Transitions should be used sparingly, depending on the effect you want the edit to have, such as conveying a change of time or location. Most editing, however, should not call attention to itself.

- How to place clips to the timeline, and add transitions and text graphics in both Final Cut Express and Adobe Premier elements.

VOCABULARY *Review*

Define the following terms:		
Capture	HDV	Rendering
Editing	In point	Time code
FireWire	Out point	Trim or Trimming
Frame accurate editing	Real time transition	USB

REVIEW *Questions*

TRUE/FALSE

Circle T if the statement is true or F if the statement is false.

T F 1. Trimming refers to eliminating parts of a single clip that you do not want.

T F 2. In points are not needed to complete an edit.

T F 3. Transitions blend frames from two clips to create a new frame.

T F 4. All transitions in Adobe Premier Elements are real time.

T F 5. Video can be imported into a Final Cut Express or APE project without capturing.

T F 6. Video with settings different from the project setting must be rendered.

T F 7. Device control means the computer can control the deck or camera functions.

T F 8. USB and FireWire are the same thing.

MULTIPLE CHOICE

1. The DV connection refers to _____.
   A. USB
   B. FireWire
   C. SCSI
   D. 9-pin connection

2. In Final Cut Express the clips you capture are stored in the _____.
   A. Browser
   B. Viewer
   C. Canvas
   D. Timeline

3. In Adobe Premier Elements the clips you will edit are stored in the _____ window.
   A. Browser
   B. Viewer
   C. Media
   D. Timeline

4. The keyboard command for undo on a Mac is _____.
   A. Command + Z
   B. Ctrl + Z
   C. Option + U
   D. Function + U

5. The keyboard command for undo on a PC is _____.
   A. Command + Z
   B. Ctrl + Z
   C. Option + U
   D. Function + U

6. The time code numbers represent _____ in order.
   A. minutes, hours, frames, seconds
   B. hours, seconds, minutes, frames
   C. seconds, hours, minutes, frames
   D. hours, minutes, seconds, frames

7. The steps to import a file into Final Cut Express using the menubar is _____.
   A. Edit > Open > Video File
   B. File > Open > Video File
   C. File > Import > Files
   D. Sequence > Import > File

8. Transitions in Adobe Premier Elements are found in the _____.
   A. File menu
   B. Viewer window
   C. Transitions window
   D. Effects window

WRITTEN QUESTIONS

Write a brief answer to the following questions:

1. Explain the editing process, no matter what editing system you are using.

2. Explain time code and its purpose.

PROJECTS

PROJECT 9-1

Tell a story with the video.

1. Create a new project file and save it as **Project 9-1** in your Digital Video Solutions folder.

2. Capture five different clips from footage you have shot using either Final Cut Express or Adobe Premier Elements. If you cannot capture video clips, then import the following five clips located in your Digital Video Data Files folder:

   Clip05.avi

   Clip09.avi

   Clip15.avi

Clip16.avi

Clip19.avi

3. Trim each clip and set an in and out point for each clip of no more than 40 seconds in length each, then add each clip to the timeline in the order you want.

4. Add four transitions: one between the first and second clip, one between the second and third clip, and one between the third and fourth clip, and one between the fourth and fifth clip in the timeline. (Experiment with different transitions)

5. Add a fade out at the end of the video after the fifth clip. (HINT: Look in the Video Transitions folder, Dissolve, then Dip to Black.)

6. Add a crawling title at the beginning of the video before the fade in transition. Try a different title template.

7. Add rolling credits at the end of the video after the fade out transition.

8. Render the video, then play the video.

9. Save the project and close the program.

 ## WEB PROJECT

Search the Web for information on how to edit a video. Look for similarities and differences on what you have learned in this chapter. Look for editing techniques that other video editors use.

 ## TEAMWORK PROJECT

Team up with another student and re-edit their video from project. Remember to tell a story. How do your edits differ from your partner? How is the story different because of the different edits?

CRITICAL *Thinking*

ACTIVITY 9-1

Watch a movie and pay attention to the number of edits that are involved in each scene. Pay attention to whether or not the edits work smoothly or call attention to themselves. How could the editing be better? How is it good? How could the editing be made better during the production process?

AUDIO EDITING

OBJECTIVES:

Upon completion of this lesson, you should be able to:

- Understand audio principles
- Add audio tracks to a Timeline
- Resynchronize audio and video
- Import audio from a CD
- Add an in and out point to audio files
- Add an audio clip to the Timeline
- Adjust audio levels
- Add keyframes to an audio track
- Add audio transitions
- Fade audio in and out

Estimated Time: 2 hours

VOCABULARY

Audio levels

Audio sweetening

Audio synchronization

Keyframes

Introduction

Sit down and watch just the credits of a big budget Hollywood movie, or go to a Web site that lists a movie's cast and crew. Notice how many people are listed for the sound department. The sound department might include a boom operator, sound mixer, music editor, sound editors, sound designer, re-recording mixer, sound effect-mixer, and foley artist . . . the list can get pretty long. Remember when we talked about how people often are more forgiving of bad video than they are of bad audio? Just look at all of the people assigned to make sure they get the audio right.

You probably do not have the luxury of having all of those people to work on just the audio for your masterpiece, but you can still get good audio. There are a couple things that you need to keep in mind in order to get good audio. First, it is hard to get good audio during post-production if you do not record good audio during production. All of the expensive, high-tech audio editing equipment in the world will not save you if you do not record decent audio. Second, give the audio priority during post-production. I edit video until it is right where I want it, and then go back and work on just the audio. In this lesson you learn how to use post-production time to get the audio right.

Understanding Audio Principles

Before you use Final Cut Express and Adobe Premiere Elements, it is important to understand a couple of the principles involved with audio. This lesson, however, covers audio during post-production and some of the basic principles you need to remember when you are trying to put your final edit together.

My friend just sent me a link to download his latest songs. He wrote the songs and recorded himself playing the guitar and singing. His friends recorded their parts, then he mixed them together. Not very interesting until you realize that my friend recorded his part in Norway; the guitarist recorded his part in Canada; the bass player recorded his part in Australia; the keyboardist recorded his part in Los Angeles; and, the drummer recorded his part in Denmark. All five of them have never been in the same room at the same time, but you would think they had been together when they recorded the songs.

Audio work is much like making a movie; it does not matter when or where things are recorded, you can edit them together and nobody will know the difference. This is made possible because of multiple audio tracks. You put the drums in one track, the bass in second track, the keyboards in a third track, and so on. You then can work on each part individually without affecting the other parts. For example, if you want the vocals to be louder than the guitar, you make the guitar track softer and the vocal track louder.

Audio for video is put together the same way. The audio in videos and movies includes not only dialog, but music and sound effects as well; imagine a romantic movie without some sappy love song, or a science fiction movie without the sounds of laser blasts and explosions. The process of adding music and sound effects to the video, as well as cleaning up and improving the audio, is called **audio sweetening**. Audio sweetening adds the final touches to any video you make.

Music can help create a mood for the film. If the music is dark and scary, the audience tenses up, waiting for something bad to happen. If you take the same visual scene and add light, fluffy, funny music, the audience is either confused, or they think something funny is going to happen. Picking the right music is a must for what you are trying to do.

Sound effects draw the audience into the experience and make it real. The audience expects to hear the sound of a motor when they see a car, and if they do not, they wonder what is going on. Sound effects also can lighten the mood, such as a zing or swoosh sound when somebody turns their head. Pick the audio elements carefully to enhance what you want to accomplish.

Working with Final Cut Express

Getting audio into Final Cut Express is simple: if you record audio when you shot your video, the audio is captured at the same time you capture your video. The audio is then added to the Timeline when you add the video. The audio automatically is added to tracks one and two. There may come a time, however, when you want the audio to go into different tracks, say tracks three and four. The default sequence settings in FCE include two video tracks and four audio tracks. That's enough audio tracks for two tracks for dialog and two for music.

Adding Audio Track to the Timeline

I have worked with a lot of beginning editors over the years and one piece of advice I have given them all is to keep things as simple as possible. One way some editors confuse themselves is by adding as many audio and video tracks as they can. That is all good until you cannot remember what audio track is where and what that video track does. It is easier for me to work with the default of four audio tracks and add new tracks only when I absolutely have to. For example, if I need to add an audio effect (like the sound of someone knocking on a door), I would look for a clear spot on a existing audio track in which to add the audio effect, before I add a new audio track for the audio effect.

STEP-BY-STEP 10.1

In this Step-by-Step you add an audio track to the Timeline.

1. Open **Final Cut Express (FCE)**, then open the data file named **SBS9-9**.

2. Click anywhere in the Timeline window to select the Timeline as the active window. The option to add tracks to the Timeline window will not be available if the Timeline window is not the active window.

3. Click **Sequence**, then click **Insert Tracks** from the menubar. The Insert Tracks dialog box appears as seen in Figure 10-1. Since you are not adding video tracks, you can ignore that part of the dialog box as long as the number in the Insert Video Tracks box is 0. If you want to make sure you do not add any unnecessary video tracks you can click on the check box next to Insert Video Tracks to deselect it.

FIGURE 10-1
Insert Tracks dialog box

4. Press **Tab**. This highlights the 0 in the Insert Audio Tracks box.

5. Type **2**. The other buttons allow you to select where the audio tracks are added to the Timeline window. If you select the Before Base Track option, the new audio tracks becomes tracks 1 and 2, and the original tracks 1 and 2 becomes tracks 3 and 4. If you select After Last Track option, the new audio tracks become tracks 5 and 6.

STEP-BY-STEP 10.1 Continued

6. Press **OK**. The two new audio tracks are added to the Timeline window. You may need to scroll down to see the added tracks. See Figure 10-2.

FIGURE 10-2
New audio tracks added to Timeline

New audio tracks added to timeline

7. Save the project as **SBS10-1**. Leave the program open for the next Step-by-Step activity.

Before we continue, let's look at the Timeline layout. On the far left you notice a round, green button, next to it a square labeled a1, and a square next to that labeled A1. The green button is called the Audible button and it determines whether or not you can hear the audio in the track. If the button is green, you will be able to hear what is in the track. If you click on the green button, it turns gray, and the track itself becomes a darker gray than before, as seen in Figure 10-3. This means the track is muted.

> **Note** ☑
>
> Sometimes the audio you want to add to your project is audio you have recorded yourself. You either can put this audio on a CD and import it, or you can record it to video tape and capture it that way. Final Cut Pro allows you to capture audio, audio and video, and video, but Final Cut Express only allows you to capture audio and video together. If all you want to use is the audio, just add the video clip (which probably will be useless video) to the Timeline, and delete the video.

FIGURE 10-3
Muting an audio track

Visible button

Muting an audio track by pressing the Audible button

The a1 and a2 squares are the audio Source buttons. Source is the audio and video you have captured. The Source buttons allow you to decide which Timeline tracks the audio (or video) will go into. The default for FCE is the left Source channel will go into track A1 in the Timeline window, and the right Resource channel will go into track A2 in the Timeline window. You can change the audio track the source files will go into by clicking and dragging the source buttons to a different destination track, such as the A3 and A4 tracks. This is helpful when you are working with a number of different tracks.

Many times you have so many different audio tracks that it is hard to figure out what audio track is making what sound. For example, not long ago I added a voice over to a video piece that had four tracks of audio. In the middle of the audio I heard a thud, but I could not figure out which track had the thud in it. I muted each audio track one at a time until I did not hear the thud. I loaded that track into the viewer, identified the sound, and then eliminated it.

Resynching Video and Audio

There is one thing that we need to cover before we go any further is audio synchronization, or audio synch. Audio is in synch when the sound matches the action. The easiest way to tell when audio is out of synch is when a character's lips move, but the sound of their voice is heard before or after the lip movement. To keep the audio linked to the video, make sure the Linked Selection button at the upper right-hand corner of the Timeline window is selected as seen in Figure 10-4.

FIGURE 10-4
Linked Selection button

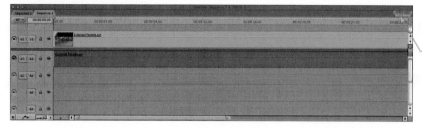

Linked
Selection
button

When the Linked Selection button is active, you can select the video in the Timeline window and the linked audio automatically is selected. You can now move the audio and video around the Timeline window as you choose. If the audio does get out of synch with the video, it is very simple to get the two back together.

S TEP-BY-STEP 10.2

In this Step-by-Step activity, you resynch out of synch audio and video. Before you resynch the audio and video, they need to be out of synch, so the first couple of steps will help you get them out of synch.

1. With **SBS10-1** open, make sure the **Linked Selection** button is deselected by clicking it. If it is deselected, the button will be grayed out.

2. Click and drag the **Italian Parade clip** to the front of the Timeline window. This adds both the audio and the video to the Timeline window.

3. Click the Italian Parade audio clip to select it. If the linked selection button has been deselected this should select the audio track only, not the video track.

STEP-BY-STEP 10.2 Continued

4. Drag the audio tracks to the left and release the mouse button. This should move the audio tracks only. A small red box will appear in the upper, left hand corner of the video track and a blue box appears in the upper, left hand corner of the audio track as seen in Figure 10-5.

FIGURE 10-5
Small red box and move into synch

Ctrl+Click
on the red
rectangle

5. **Ctrl+Click** on the red rectangle with numbers in the upper left hand corner of either the audio or video clip. A drop down menu appears.

6. Click **Move into Synch** and the audio and video move back into synch.

7. Close the project. Do not save the file. If possible, leave the program open for the next Step-by-Step activity.

Import an Audio File From a CD

You are now going to import some audio into FCE. You can grab your favorite CD, or pick an mp3 file. The advantage of picking a song from a CD is that the music file does not need rendering in FCE, while an mp3 or similar type of file needs rendering before you can be work with it.

STEP-BY-STEP 10.3

In this Step-by-Step you import an audio file from a CD. The steps apply to importing any type of audio file. If you want to import an mp3 or you can not import an audio file from a CD, start with step 4.

1. With **SBS10-1** open, insert the CD into the CDROM drive. This may open iTunes, but do not worry, you can close iTunes by clicking iTunes on the menubar, then click **Quit iTunes**. The CD appears on the desktop. You may need to move the Canvas and Timeline window window to the left by clicking the Canvas title bar and dragging it to the left.

2. Double-click the CD icon on the desktop to open it. This shows you all of the files on the CD. This step is important if you are importing audio from a CD.

3. Click the file you want and drag it to the desktop. This copies the file to the desktop. Close the CD window.

4. Launch FCE if it is not running already.

STEP-BY-STEP 10.3 Continued

5. Click **File**, point to **Import**, and then click **Files** from the menubar to open the Choose a File dialog box.

6. Click **Desktop** icon on the left side of the Choose a File dialog box to get to the file you copied to the desktop. If you did not do steps 1-3, then click on your data file folder.

7. Click the name of the audio file you want to import. If you are using a data file, click **Music file 1.aiff**.

8. Click **Choose** to import the file and place it in the FCE browser.

9. Save the project as **SBS10-3** and leave the project for the next Step-by-Step activity.

> **Note**
>
> If you import the audio file directly from the CD, FCE links to the CD, which means that the audio link will be lost when you take the CD out of the computer.

Adding Audio In and Out Points

Sometimes, however, you don't want to add the entire audio clip to the Timeline window. Say, for example, that you have a nice piece of music, but the very beginning has some weird sounds that you just do not want in your video. All you have to do is create an in and out point, just like you did for video clips, and add the clip to the Timeline window.

STEP-BY-STEP 10.4

In this Step-by-Step activity, you add audio in and out points to a clip.

1. With **SBS10-3** open, drag the audio clip you just imported from the Browser window to the Viewer window.

2. Move the playhead to the place where you want the audio clip to start. You can do this by playing the audio and listening until you reach the place where you want the audio clip to start or by using the right and left keys on the keyboard.

3. Click the **Mark In** button at the bottom of the Viewer window.

4. Repeat step 2 to locate the out point, or the place where you want the audio to stop.

5. Click the **Mark Out** button at the bottom of the Viewer window.

6. Save the project as **SBS10-4** and leave FCE open to complete the next Step-by-Step activity.

Add an Audio Clip to the Timeline

Once you add the audio file to the browser it is a project file you can use, but it must be added to the Timeline window in order for it to be part of the edit. Adding audio files is the same as adding video or other files to the Timeline window. The audio clip you add to the Timeline window will include the in and out points you created in Step-by-Step 10.4.

STEP-BY-STEP 10.5

In this Step-by-Step you add audio file to the Timeline.

1. With **SBS10-4** open, click and drag either your audio file or **Music file 1.aiff** from the Browser window to A3 track on the Timeline window.

2. Click the **a1** tab next to **A1** and drag it down next to **A3**. This means that the left channel of the audio clip will be added to track A3.

3. Click the **a2** tab next to **A2** and drag it down next to **A4**. This means that the right channel of the audio clip will be added to track A4 as seen in Figure 10-6.

FIGURE 10-6
Add audio to Timeline

Adding
audio and
changing
the source
and
destination
buttons

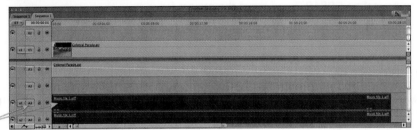

4. Save the project as **SBS10-5** and leave the file open for the next Step-by-Step activity.

Adjusting Audio Levels

Audio tracks that play at the same time are heard at the same time, so you can hear audio from tracks 1 through 99 at the same time. For example, in audio tracks 1 and 2, you could have the dialog and music; in tracks 3 and 4, you could have a dog barking; in track 5, a car driving by and, in track 6, and the sound of the people walking by. The possibilities are endless. The problem might be, however, that there is way too much audio going on at the same time that the audience is unable to separate the sounds from each other. One way to avoid this is to adjust the **audio levels**. Adjusting the audio levels means to change the volume for each track. Some audio you want to be louder, like the dialog, and others you want quieter, like the sound of the dog barking in the background. Things further away are, of course, not as loud as things that are closer.

STEP-BY-STEP 10.6

In this Step-by-Step activity, you adjust the audio levels. There are a couple of different ways to adjust in FCE: in the Viewer window and directly in the Timeline window. You cover both in this activity.

1. With **SBS10-5** open, double-click either your audio file or **Music file.aiff** in the Timeline window to load it into the Viewer window, if it is not loaded already in the Viewer window. If you do not see the audio file in the Viewer window, then click the Stereo tab located at the top of the Viewer Window.

STEP-BY-STEP 10.6 Continued

2. Click the **level slider** at the top of the Viewer window as seen in Figure 10-7, and drag it slowly to the right.

FIGURE 10-7
Level Slider on the Viewer window

Level slider

The level starts at 0 when you load the audio, and moves into positive numbers. Positive numbers means the audio level is higher (louder). Dragging the slider to the left lowers the volume or level, which is signified by negative numbers. The audio level in the Timeline reflects the level adjustments made in the Viewer window.

3. To adjust the audio level directly in the Timeline window, click the **Toggle Clip Overlays** button on the lower left hand corner of the Timeline window. A pink line appears in the middle of the audio clip in the Timeline window as seen in Figure 10-8.

FIGURE 10-8
Pink line in the middle of the audio clip

Pink line in middle of the audio clip

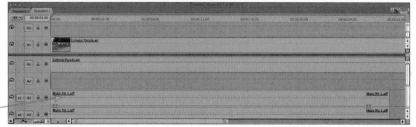

4. Move the mouse pointer over the pink line. The pointer becomes an icon with one arrow pointing up and another pointing down.

5. Click and drag the line up to raise the volume, and drag it down to lower the volume.

6. Double-click the **audio clip** on the Timeline window to load it into the Viewer window. You can see the changes you make in the Viewer window.

7. To figure out how much to raise or lower the audio level, click the **Play** button on the Viewer window.

8. While the audio clip is playing, you can move the pink line on the audio clip either on the Timeline window or on the Viewer window to adjust the volume.

> **Note**
>
> I usually set my playback level (speaker volume) somewhere in the middle. If the speaker volume is too high you may set the audio too low, and if it is too low, you might adjust the audio too high.

STEP-BY-STEP 10.6 Continued

9. Another way to check the volume is to check the audio meter. The audio meter is the long, skinny, colorful window to the right of the Timeline window as seen in Figure 10-9.

The level moves up and down when you play the audio. The main audio should be bouncing between –12 and just above the –6, or between the lower yellow and just into the red, but background sounds shouldn't be as loud. One thing to avoid is for the red circle at the top of the level meter to light up. This means that audio is too loud and will be distorted on playback.

> **Note** ☑
>
> If you don't see the audio meter, click **Window** on the menubar. If there is not a checkmark next to the Audio Meters then click **Audio Meters** option.

FIGURE 10-9
Audio Meter

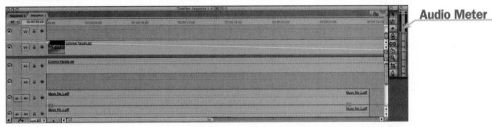

Audio Meter

10. Click the Play button on the Viewer window and watch the audio meter. Click **Play** again to stop the audio.

11. Save the file as **SBS10-6** and, if possible, leave the file open for the next Step-by-Step activity.

Adding Keyframes to the Audio Level Overlay

Sometimes you only want to adjust the audio level for a segment of the clip, not the entire clip. This is important when you have a mistake in the audio, like a thump or something that you need to get rid of so the audience does not hear it. This is done by adding keyframes to the track. Keyframes identify a specific position, in this case a specific frame, in the Timeline to start and end a level change.

S TEP-BY-STEP 10.7

In this Step-by-Step you add keyframes to the audio level overlay.

1. Make sure the file named **SBS10-6** is opened. If the pink audio level overlay line does not appear in the Timeline window, click the **Toggle Clip Overlays** button at the bottom left hand corner of the Timeline window.

2. Click the Pen Tool at the bottom of the Tool Palette as seen in Figure 10-10.

STEP-BY-STEP 10.7 Continued

FIGURE 10-10
Pen Tool

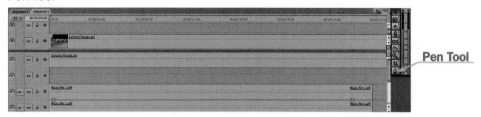

Pen Tool

3. Click the pink audio level overlay line on either audio clip or **Music file.aiff** in A3 track of the Timeline window with the Pen Tool. A diamond shape will appear on the overlay as seen in Figure 10-11.

FIGURE 10-11
Diamond shape on the pink audio level overlay line

Diamond shape (keyframe) on pink audio level overlay line

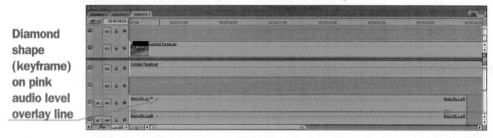

4. Repeat step 3 at a different location on the Timeline window to create a second keyframe.

5. Create two other keyframes on the Timeline window by repeating step 3. The Timeline window should look like Figure 10-12.

FIGURE 10-12
Adding keyframes to theTimeline

Adding keyframes

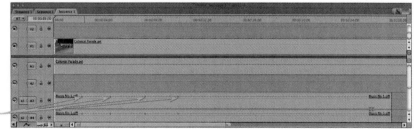

6. Move the mouse over the second diamond from the left. The cursor should appear as a plus sign or crosshair.

7. Click and hold the cursor on the diamond and drag it down to –26 dB to lower the audio volume at that point in the Timeline window.

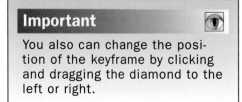

Important

You also can change the position of the keyframe by clicking and dragging the diamond to the left or right.

STEP-BY-STEP 10.7 Continued

8. Repeat step 7 for the third keyframe and move the diamond to –8dB. This changes the volume between second and third keyframes. The first and fourth keyframes serve as anchors so the volume for the rest of the clip remains where it was before.

9. Make sure the playhead is at the beginning of the audio clip in the Timeline window or in the Viewer window, then click **Play** on the Viewer window and listen to the audio changes.

10. Save the file as **SBS10-7** and leave the file open for the next Step-by-Step activity.

Adding an Audio Transition

Often one audio clip does not mix well with the audio clip following it, or you need some way to smooth out the transition between silence, like at the beginning of the video and the first of the audio. Audio transitions are added in the same way that video transitions are added. A transition can be placed between two audio clips, or between silence (no audio in the track) and an audio clip.

STEP-BY-STEP 10.8

In this Step-by-Step you add an audio transition between two audio clips.

1. With the **SBS10-7** open, import another audio clip of your own or **Sound file 1.aiff** to the Browser window. Add in and out points to the audio clip, then add it next to the audio clip in the A3 track on the Timeline window.

2. Move the playhead on the Timeline window so it is between the two audio clips as seen in Figure 10-13.

You can click the playhead and drag it to the beginning or you can press the ; key until you reach the beginning of the clip.

FIGURE 10-13
Positioning the playhead for the transition

Position
playhead
between
the two
clips

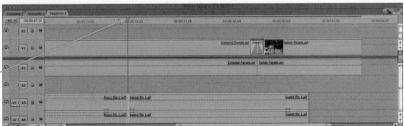

3. Click the **Effects tab** in the Browser window.

4. Click the triangle to the left of the **Audio Transitions folder**.

5. Click and drag the **Cross Fade (0dB)** icon to the position of the playhead between the two clips as seen in Figure 10-14.

STEP-BY-STEP 10.8 Continued

FIGURE 10-14
Adding a transition to the Timeline

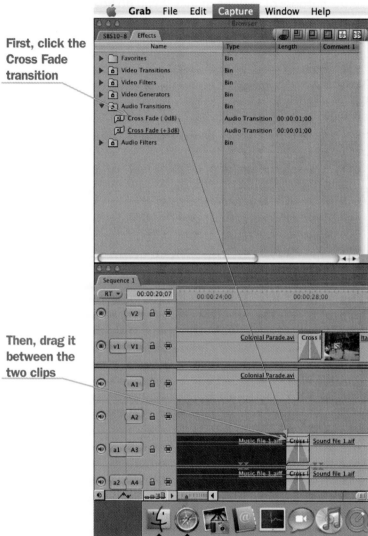

First, click the Cross Fade transition

Then, drag it between the two clips

6. Make sure the playhead is at the beginning of the audio clip on the Timeline window, then click **Play** to listen to the transition.

7. Save the project as **SBS10-8**, close the project, and exit FCE.

Working in Adobe Premiere Elements

Importing Audio

You can import a number of different audio files into Premiere Elements, but you have to be aware of a couple of issues. First, some downloaded music from iTunes or another online music source, cannot be played back in Premiere Elements. Second, music from a CD will need to be ripped using ripping software. The most accessible ripping software already should be on your Windows machine—Windows Media Player. Remember to respect copyright. Copyright means a person has the right to legally reproduce, sell, or distribute creative or intellectual materials. You cannot use music or images in a video for the purpose of making money if you do not own the copyright or have written permission from the copyright holder. However, there is an exemption to copyright called Fair Use. Teachers are granted access to works for use in the classroom and for educational purposes. There are four considerations for how to determine if copying materials falls under Fair Use:

- **Purpose of use.** Selected parts are used for educational purposes only. Copies are made spontaneously, meaning that if you want to show an interview that appeared on *60 Minutes* last night for your class the following day, this is covered under fair use. However, if you want to use that same interview for your class next year, this is not covered under fair use. You must legally obtain the material by purchasing it. You cannot redistribute.

- **Proportion of materials used.** If the duplication is short in relation to the entire work, this falls under fair use. The general rule is 10%, unless a maximum amount is set. Segments do not reflect the essence of the work. The essence of the work could be the climax of the work, for example, how a story ends or what happens to a main character or an idea.

- **Nature of work.** You may copy parts of material that do not reflect the essence of the work. Facts, names, ideas, public images are fair use.

- **Effect on marketability.** Use of the work should not cause the reduction in sales of the work.

Fair use guidelines can be found on the Web: A Teacher's Guide to Fair Use and Copyright *http://home.earthlink.net/~cnew/research.htm*

STEP-BY-STEP 10.9

In this Step-by-Step you rip a song from a CD using Windows Media Player. This is the same as copying the file from the CD to the computer. Remember, respect copyright. You also import the song so it can be added to the Timeline window and included in the edit. So far you copied the audio clip to the computer, but Premiere Elements does not know that you want to include it in the video until you have imported it into the project file. If you can not rip a song from a CD, go to step 6 and use the data file indicated.

1. Insert a music CD into the computer. If Windows Media Player 11 does not launch automatically, click Start, then point to **All Programs**, and then click **Windows Media Player**. Windows Media Player opens the library, showing you the audio already in your library.

2. Click **Rip** at the top of the screen. This opens the ripping window, showing you the tracks that can be ripped.

3. All of the tracks that have not been ripped and are available to be ripped have a checkmark in the box to the left of the song title. Click inside the box to remove the checkmark for the songs that you do not want to import. Select two songs.

4. Once you have selected two songs you want to rip, click **Start Rip** in the bottom right hand side of the window. The music is added to the library and stored in the **My Music** folder in **My Documents** on your hard drive or network drive, unless you have set a different location for your music to be stored.

STEP-BY-STEP 10.9 Continued

5. Close **Window Media Player**. The song you just ripped can now be imported into the Premiere Elements and added to the Timeline window.

6. Open Adobe Premiere Elements 2.0. When the open screen appears, click **Open Project** and find the file named **SBS9-15.prel**.

7. Click **Add Media** at the top left hand side of the screen.

8. Click **From Files or Folders**. This opens up the Add Media dialog box.

9. If you did not change the rip destination, Double-click **My Documents** from the left hand side of the window. Click **My Music** folder from the list. This opens the folder and give you a list of folders and files. If you are using a data file, click your Data file folder.

10. Find the song you want or use the data file **Music file 1.aiff**. Click the song, then click **Open**. The song is placed in the media panel, ready to be added to the Timeline window.

11. Import another audio file following steps 7 through 10. Use the data file **Sound file 1.aiff**.

> **Note**
>
> Depending on how you have set up your music, you may see the song you want in that folder, or you may need to open other folders to find what you are looking for. For example, if the name of the band that you ripped the music from is Joe's Band, you may see a folder named Joe's Band. If you have ripped music from more than one of Joe's Band's CDs, there may be other folders with the names of CDs.

12. Save the project as **SBS10-9** and leave the project open for the next Step-by-Step activity.

Adding In and Out Points to the Audio Clip and Adding the Clip to the Timeline

The audio can now be added to the Timeline. Sometimes, you only want to add specific pieces of the audio to the Timeline. For example, if the car engine sound effect includes the car starting, but all you want is the sound of the engine after you start the car, then you set an in point right after the engine starts and you eliminate the car starting sound. Audio in and out points are set just like video in and out points.

STEP-BY-STEP 10.10

In this Step-by-Step you add in and out points to the audio clip.

1. With **SBS10-9** open, click in the Media panel to select it.

2. Double-click your first audio clip or **Music file 1.aiff** to load it into the monitor window.

3. Click the **Play** button on the Monitor panel to play the clip.

> **Note**
>
> You can be even more precise with the placement of the in point (or out point for the next step) by using the right and left arrow keys on the keyboard to move through the clip.

STEP-BY-STEP 10.10 Continued

4. When you hear the start of the audio you want, click the **Set In Point** button on the Monitor panel.

5. Click the **Set Out Point** button to set the audio out point about 25 seconds into the clip to end the clip. The audio now has an in and an out point and can now be added to the Timeline window.

6. Click the clip in the Monitor panel and drag it to the beginning of Audio 2 track in the Timeline window as seen in Figure 10-15. The clip is added to the Timeline window.

FIGURE 10-15
Clip added to the Timeline

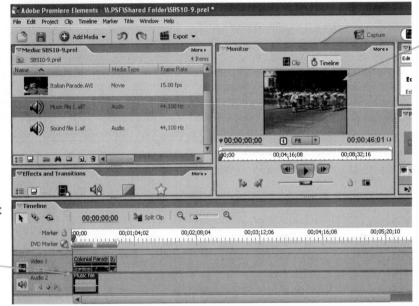

First, click the clip in the Monitor panel

Then, drag it to Audio 2 track on Timeline

7. Repeat steps 2 through 5 to add in and out points to the second clip or **Sound file 1.aiff** (about a 45-second clip).

8. Repeat steps 6 to add your clip or **Sound file 1.aiff** to the Timeline window after the first audio clip. Your project file should look like Figure 10-16.

FIGURE 10-16
Adding audio clips to Timeline

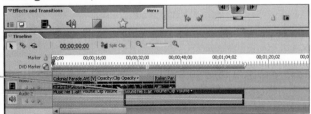

Two audio clips added to timeline

9. Save the project as **SBS10-10**. If possible, leave the project open for the next Step-by-Step activity.

Adjusting the Volume From the Timeline

Now that the audio clip is in the Timeline you can adjust the volume. There are a couple of ways you can do this, so you can decide which way you want to do it. Premiere Elements handles audio different than FCE.

First, audio with two tracks is imported into a single audio track. For example, you import an mp3 audio file with a left and right channel. Both the left and right channels are imported into the same audio track. (Channels can be thought of in the same way as you think of tracks in either FCE or Premiere Elements.) This means it is more difficult to work with each track individually.

Audio in more than one channel allows you to work with one channel at a time. For example, you record two people talking, and the second person starts talking before the first person is finished. You really want to hear what the first person says, but the second person has interrupted. Recording each speaker in a separate channel allows you to turn one channel off so you can hear the other channel. When I shoot an interview I have the interviewer in one channel, and the interviewee in the other channel so I can adjust one without having to worry about the other. This may, or may not, be a problem, depending on what you are doing.

Second, you can not adjust anything in an audio clip until the clip is added to the Timeline window.

STEP-BY-STEP 10.11

In this Step-by-Step you adjust the audio volume from the Timeline window.

1. With **SBS10-10** open, click and drag the yellow line in the middle of your clip to or **Music file 1.aiff** adjust the audio level. Dragging the line down lowers the audio level (volume), dragging the line up raises the audio level (volume).

2. Drag the yellow line on the first clip or **Music file 1.aiff** to **–5.37 dB**. Drag the yellow line on the second clip to **5.67 dB**.

3. Make sure the playhead is at the beginning of the clips. Click the **Play** button on the Monitor panel.

4. Adjust the audio overlay lines as you listen.

5. Save the file as **SBS10-11** and leave the project open for the next Step-by-Step activity.

Create Keyframes and Adjust Audio Level

Sometimes, you want to adjust the level for a section of a clip, and not the entire clip. You may hear a bump that you want to eliminate, or have a situation where one line in the dialog isn't as loud as you want it. You can raise or lower a specific section by adding keyframes to the clip.

STEP-BY-STEP 10.12

In this Step-by-Step you create keyframes and adjust the audio level.

1. With **SBS10-11** open, position the playhead at a point a couple of seconds into the first audio clip or **Music file 1.aiff** that you placed on the Timeline window. You can move the playhead by either using the mouse to click and drag the playhead, or use the right or left arrows on the keyboard.

2. Click the audio clip to select it.

3. Click the **Add/Remove Keyframe** button. The Add/Remove Keyframe button is located between the two arrow points to the left of the audio track. This adds a diamond shape along on the yellow line in the audio clip as seen in Figure 10-17.

4. Position the playhead to another position in the Timeline window using either the mouse or the keyboard arrow keys.

5. Create a second keyframe by clicking the **Add/Remove Keyframe** button. You can now adjust the audio level by clicking one of the keyframes and dragging it up or down. Dragging it up increases the volume, and dragging it down lowers the volume as seen in Figure 10-17.

FIGURE 10-17
Adding keyframes and adjusting the volume

Adding
keyframes
to timeline
and adjusting
the volume

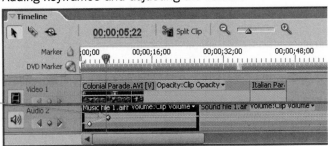

6. Play the clip to hear the adjusted volume levels.

7. Save the project as **SBS10-12** and leave the project open for the next Step-by-Step activity.

Add a Transition

You need to add a transition. An audio transition is added to the Timeline window in the same way a video transition is added.

STEP-BY-STEP 10.13

1. With **SBS 10-12** open, make sure that you have at least two audio clips in the Timeline window.

2. In the **Effects and Transitions** panel, click the arrow next to the **Audio Transitions folder** icon. You may need to scroll down to see the folder. Click the arrow next to **Crossfade** as seen in Figure 10-18.

STEP-BY-STEP 10.13 Continued

FIGURE 10-18
Cross Fade transition

Crossfade Constant Power icon

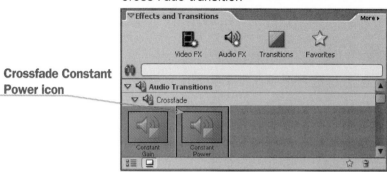

3. Click the **Crossfade Constant Power** icon and drag it to the seam between the two clips. Release the mouse button to add the transition to the Timeline window.

4. To preview the transition, use the arrow buttons or click and drag the playhead over the transition.

5. Save the project as **SBS10-13** and leave the project open for the next Step-by-Step activity.

Fade Audio In and Out

There is one thing to be aware of before we end this lesson. Audio fades in FCE are handled by the same effect as any other cross dissolve, but that is not so in Premiere Elements. A fade in or fade out is added to the clip through the Properties window.

STEP-BY-STEP 10.14

In this Step-by-Step you fade the audio in at the beginning of the audio clip and fade the audio out at the end of an audio clip.

1. With **SBS10-13** open, double-click the first audio clip or **Music file 1.aiff** on Audio 2 track in the Timeline window to load it into the Monitor panel.

STEP-BY-STEP 10.14 Continued

2. Click the triangle to the left of Volume in the **Properties** panel. A drop down menu appears with Clip Volume, Fade In, and Fade Out as seen in Figure 10-19.

FIGURE 10-19
Properties Panel

3. Click **Fade In** if the audio is the first clip in the Timeline window. Notice the audio line at the beginning of the first clip. See Figure 10-20.

4. Click the second video clip. Click **Fade Out** in the **Properties** panel. Notice the audio line at the end of the second clip as seen in Figure 10-20.

FIGURE 10-20
Fade out on the audio clip

Fade In on the audio clip

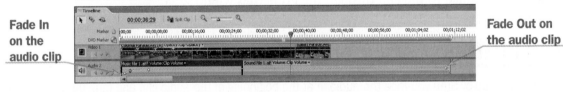

Fade Out on the audio clip

5. Click **Timeline** then click **Render Work Area** to render the video in the Timeline window.

6. Play the clip.

7. Save the project file as **SBS10-14**.Close the application.

Those are the basics of working with audio in Final Cut Express and Adobe Premiere Elements. Get the basics down, then experiment with the other features.

SUMMARY

In this lesson you learned:

■ Finished audio includes dialog, music, and special effects. It is possible to work with each audio element by loading each one into an audio track so each can be worked with separately.

■ You can add additional audio tracks to a Timeline in order to add the separate audio elements.

■ Audio files can be imported into Final Cut Express and Adobe Premiere Elements.

■ The audio level of audio clips can be adjusted in order to make the audio volume louder or softer. Levels can be adjusted for the entire clip, or adjusted for only segments of the clip.

■ Sections of the clip can be adjusted by the use of keyframes.

■ Transitions can be used to smooth the change between audio clips.

■ Audio also can fade in and fade out. An audio fade in and fade out is a normal transition in Final Cut, but is added through the properties window in Premiere Elements.

VOCABULARY *Review*

Define the following terms:

Audio levels   Audio synchronization   Keyframes
Audio sweetening

REVIEW *Questions*

MULTIPLE CHOICE

Select the best response for the following questions:

1. Which of the following menubar menus adds audio tracks to the Timeline in Final Cut Express?
   A. click **Sequence**, then **Insert Tracks**
   B. click **Sequence**, point to **Settings** , then click **Add Audio Tracks**
   C. click **Modify**, point to **Sequence**, then **Settings**, then click **Add Audio Tracks**
   D. click **Sequence**, then click **Add Tracks**.

2. Which audio transition in Premiere Elements is added from the Properties window?
   A. wipe
   B. fade in
   C. dissolve
   D. page turn

3. Audio transitions
   A. fix all sorts of problems
   B. add emotion to the audio
   C. smooth two audio clips together
   D. are added during production

4. Which of the following statements describes what music can do for the audio?
   A. adds emotion to the video
   B. slows the pace of a video
   C. distracts the viewer from the video
   D. should be added to the same audio track as the dialog

5. Which of the following statements describes the Audio levels in FCE?
   A. are adjusted in the Browser window
   B. cannot be adjusted once they are added into the Timeline
   C. can be adjusted in the Timeline window and the Viewer window
   D. must always be adjusted

6. In Premiere Elements, which of the following options is used to add a keyframe to the Timeline?
   A. clicking **Timeline**, then **Add Keyframes** from the menubar
   B. clicking **Transition**, point to **Audio**, then **Transitions**, then clicking **Keyframes** from the Effect and Transitions panel
   C. clicking the **Pen Tool** in the Tool menu
   D. clicking the **Add/Remove Keyframes** button

7. Which of the following options describes what Keyframes allow you to do?
   A. adjust the level for an entire clip
   B. synch the audio with the video
   C. adjust specific sections of an audio clip
   D. identify transition points

8. In Premiere Elements, how are audio clips added?
   A. to the media panel
   B. to the audio panel
   C. directly to the Timeline
   D. at −50 dB

TRUE/FALSE

Circle T if the statement is true or F if the statement is false.

T   F   1. Good video makes up for bad audio.

T   F   2. Any audio problem can be fixed during post-production.

T   F   3. Audio transitions can smooth the changes between two audio clips.

T   F   4. Audio levels can be changed in the Timeline in FCE, but not in Premiere Elements.

T   F   5. Keyframes can be used to control audio levels in specific segments of the video.

T   F   6. Sweetened audio can include dialog, music, and special effects.

T   F   7. An audio level can only be adjusted for the entire clip—sections of a clip cannot be altered.

T   F   8. The volume, or audio level, is measured in volume units

WRITTEN QUESTIONS

Write a brief answer to the following questions.

1. Explain audio keyframes and give an example of when you would want to use them.

2. Describe Synchronization and give an example, either of when audio is in synch or out of synch.

3. Explain audio sweetening and how it helps the video.

PROJECTS

PROJECT 10-1

1. Open Final Cut Express or Adobe Premiere Elements.

2. Open Project 9-1.

3. Adjust the audio levels for each clip so they sound the same (if you adjust the audio levels to the same level, some may actually be louder than others).

4. Add a music track to the Timeline and adjust the level so it can be heard, but does not distract from the other audio. The music should be appropriate for the mood and emotion you are trying to create.

5. Add special effects where necessary. You can find a number of special effects on the Internet, or record and capture your own.

6. Complete the audio so that you have a final edit and final video to show and audience. Add transitions or fades, or whatever you need to complete the show.

7. Save the project as **Project 10-1** and close the project.

PROJECT 10-2

1. Open Final Cut Express or Adobe Premiere Elements.

2. Capture or import at least two audio clips that are not associated with a video clip you have already captured.

3. Set an in and out point for at least one of the clips.

4. Adjust the audio level for one entire clip.

5. Use keyframes to adjust the level of the second clip.

6. Add a transition between the two clips.

7. Add a fade in at the beginning of the first clip.

8. Add a fade out at the end of the second clip.

9. Save the project as **Project 10-2** and close the project.

 WEB PROJECT

Watch the credits at the end of a movie, or go to a Web site like *www.imdb.com* and identify the different titles of the crew members involved with the audio. Some of those titles are listed at the beginning of this chapter. Research what each position does (*Hint*: *www.imdb.com* has a glossary about all of those different jobs) and explain in your own words (on paper) the importance of each crew member.

 TEAMWORK PROJECT

Get with at least one classmate and record a reading from part of a favorite book. First, add music to the recording and talk about how the choice of music helped or hurt the passage. Then, see if you can change the meaning of the passage by changing the order of words by editing the audio track (don't read it differently, cut it up and move words around). How does the audio sound now? Does it sound jumpy or smooth? This is great practice for working with voice over material.

CRITICAL*Thinking*

ACTIVITY 10-1

Rent a video or watch a television show (but not the news or a talk show). Pay specific attention to the audio, particularly the sound effects and music. How does the music enhance the experience? Does the music create an appropriate mood, or does it take away from the mood? Would the show be better with different music? What would the experience be like without the sound effects?

ACTIVITY 10-2

Experiment with the AudioDJ software for your own project. (Perform a Google search for AudioDJ to find out more about the software and how to get a trial copy) Add odd or goofy sound effects, or out of place sound effects to the audio track. How does it change the mood of the video? Does it help or hurt what you are trying to do? Play around with sounds and music and have fun.

DELIVERY

Introduction

By this point you have written a script, as well as scheduled, shot, and edited your video. Now it is time for you to show your movie. You need to deliver the video in a format that people can see.

You could have everybody watch it on your computer, but how is Aunt Nellie in Kalamazoo going to see it when your computer is in Boise? You do not want to package up your computer and send it through the mail. You could put it on one of many tape formats, but there is no guarantee that whoever you send it to has the right playback deck.

You need to create a video file to send to family and friends. The problem is that the video you just finished is probably about 12 gigabytes per hour of video. That means that if your video is a half an hour long you need at least 2 DVDs to put it on. You need to figure out a way to make that video file smaller, but still viewable. In this lesson, you learn about compression basics.

Compression Basics

Let's say you build a really cool car with colored, interlocking blocks. You think it is so cool that you want your friend to have one too, but he lives on the other side of the world. You could send him the car in the mail, but you do not have enough money. Besides, you want your blocks back and he probably will not send them back. You decide to write careful instructions so your friend can build the car with his own interlocking building blocks. You send the instructions to him with an example of one of each of the blocks he needs. For example, he needs 12 identical blue blocks and 12 identical red blocks. You send him one of the blue blocks and one of the red blocks so he knows exactly which blue and red blocks to use. That saves you money and blocks and with the perfect instructions he can build an identical car.

Video compression follows that same idea with pixels as the building blocks and a codec as the instructions. Codec is short for compression/decompression. A video codec is a complicated mathematical **algorithm** (or set of instructions) that looks at each frame of a video and determines the best way to handle the redundant (or repeated) information during the compression phase, then determines how to reconstruct the information during the decompression phase. The decompression phase is when somebody watches the video.

Some of the most common codecs are Sorenson, H.264, MPEG-1, MPEG-2, and MPEG-4, but there are many, many more. It can get rather confusing when talking about codecs because of what I call the "wrapper issue." The wrapper simply tells QuickTime that there is a file inside that it can display. Apple created QuickTime to play media objects on the computer. Files that play in QuickTime have the file extension .mov. The file extension .mov identifies the wrapper, but not the codec. A video compressed to play in QuickTime using H.264 will have a .mov extension, but so does a video compressed using the Sorenson codec. The same can be said of Window Media Player's wrapper, .wmv.

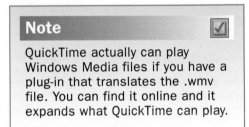

Note ☑️

QuickTime actually can play Windows Media files if you have a plug-in that translates the .wmv file. You can find it online and it expands what QuickTime can play.

All codecs come in one of two types, spatial and temporal. A **spatial codec** compares each pixel in a frame to all of the other pixels in that frame (remember, there are 30 still frames per second in video). **Temporal codecs** compare the pixels in one frame to the pixels in the other frames.

Let's say you have a one second video of a clear blue sky in which every pixel in each frame is exactly the same. A spatial codec examines frame one and decides whether every pixel in the frame is identical to every other pixel in the frame. It then saves one pixel from frame one and leaves instructions to copy that pixel enough times to re-create the frame during playback. This happens for each of the 30 frames, saving one pixel and instructions to copy that pixel enough times to recreate the frame during playback.

This example is an oversimplification because not every pixel in a frame is going to be exactly like all of the other pixels in the frame. The more complex the image in each frame (meaning more color, more detail, and more information) the greater the amount of data the codec has to save. A spatial codec generally requires lots of storage space because lots of data has to be stored for each frame.

A temporal codec saves the entire first frame (not just a pixel or two). The codec then looks at the second frame and recognizes that the second frame was the same pixel information as the first frame and leaves instructions to copy frame one when it comes time to show frame two. The instruction for each frame after that is basically "copy the previous frame."

Again, this is an oversimplification, but it shows the basic difference between the two types of video compression. Where a spatial codec stores information and instructions for each frame, a temporal codec can store information and instructions for just one frame. The rest of this lesson is spent discussing temporal compression because it is the type of compression used for the delivery methods covered here, mainly Web and DVD delivery.

I mentioned earlier that a temporal codec stores the first frame. That first frame is called a keyframe by some codecs, and an I-frame by others. A **keyframe** is a complete frame or picture with all of the data needed to display that frame. For example, if the video is of a red ball moving from left to right across a blue sky, the keyframe is the red ball and the blue sky. If you see just the keyframe, you see a picture of a red ball with a blue sky.

If you look at the second frame, however, all you see is the red ball a little further away from the left side of the frame. The third frame is like the second frame, but with the ball in a different place. The fourth frame is based on the third frame but with the ball in a different place. The instructions in each frame following the keyframe are basically "copy the previous frame, except for where the ball is." See Figure 11-1. These non-keyframe frames are called **delta frames** in some codecs, and **P frames** (for predictor frames) in others.

FIGURE 11-1
Delta frames

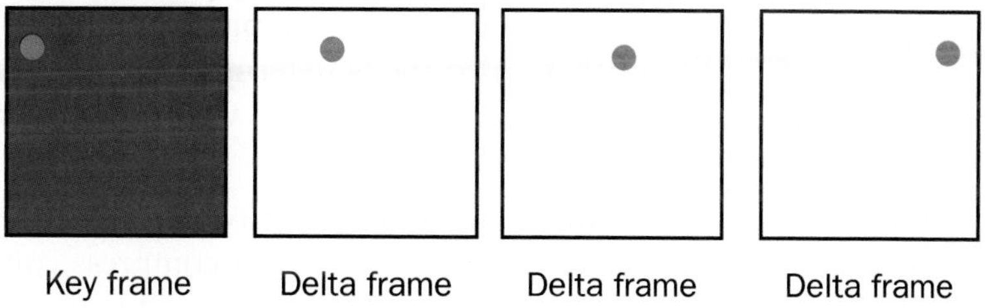

| Key frame | Delta frame | Delta frame | Delta frame |

If the scene changes—say it goes from a red ball on a blue sky to a red ball on green grass—a new keyframe is needed. Most codecs detect the amount of change the first time they see the different background (in this example the green grass instead of blue sky) and automatically creates a new keyframe. The delta frames following the second keyframe copy the second keyframe and put green grass behind the red ball (see Figure 11-2).

FIGURE 11-2
Key frames and delta frames

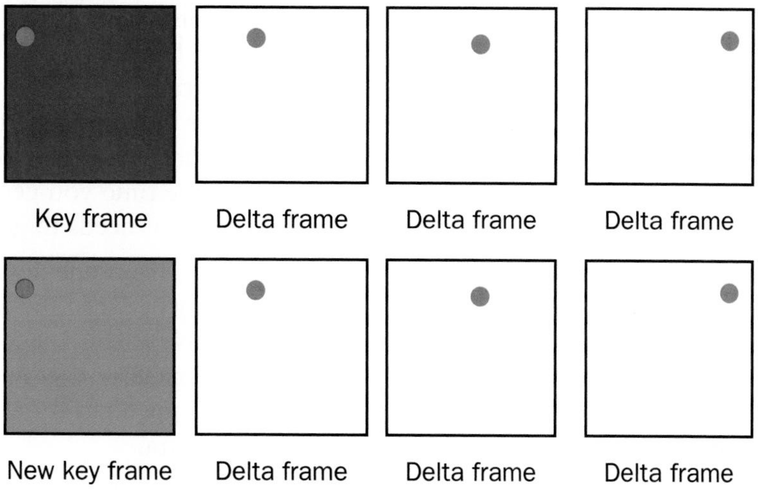

| Key frame | Delta frame | Delta frame | Delta frame |
| New key frame | Delta frame | Delta frame | Delta frame |

The problem with delta frames is that it is hard to access individual delta frames. Frame four, for example, does not really exist; only the changes exist. The computer has to make frame two from the keyframe, and then make frame three before it can make frame four. If you watch the video it is just fine, but if you want to go to one particular frame that is not a keyframe it can take a lot of time. One way to get around this problem is to tell the computer to create frequent keyframes. If you have a keyframe every 30 frames, then you would have 60 keyframes for each minute of video. A keyframe every 120 frames would create a keyframe every 4 seconds, giving you 20 keyframes each minute. Frequent keyframes, however, require more storage space.

Some codecs add a third type of frame called a **bi-directional frame**. These frames are like delta frames in that they not only track the changes, but they compare themselves to the previous keyframe and to the delta frame after it to decide what they look like.

Delta frames, or P frames, along with bi-directional frames, make it difficult to edit video using a temporal codec. Frame accurate editing requires frames. If a frame does not really exist, you can not access it to cut it out. DV is a spatial codec with each frame compressed, but it is independent of the frames before and after it.

Compression Settings

It is important to realize that what you are doing is creating a video file. A video file is just like any other file on your computer: you can copy it, you can put it on a CD, or you can send it in an e-mail. The file size and quality of that video file is determined by the settings you use. In this part of compression basics you learn to determine how the video file looks and how big the file will be.

To start off, compression tools generally come with preset options. The people that created the compression software created these presets so anybody can compress some decent looking video without having to know much about the process. For example, if you want to deliver video so that someone with a dial-up Internet connection can watch, simply select the dial-up preset. If you want to put the video on a CD, you simply select the CD preset. I still use the presets as a starting point. If I have a project that needs to go on a CD, I use the CD preset, and then take a look at it. If the video looks good I go with it, but if I do not like it, I make adjustments based on the preset to improve the compression.

The **Compression settings** define the type of codec to be used for both the audio and video, the data rate for the audio and video, and the frame size. Each element is discussed briefly here.

The main compression element is the data rate. We talked about data rates back in Lesson 1, but we need to review them again before you start compressing. The data rate refers to how much information is transferred from one device to another each second. Think about data in a computer like a bunch of delivery trucks on the freeway. Each truck has items it needs to deliver. Some trucks deliver operating system data, other trucks deliver video information, other trucks deliver e-mail, and so on. As long as each truck gets to its destination on time the computer runs smoothly. When too many trucks are on the freeway at the same time you get traffic jams. The computer freezes up because the information is not getting where it needs to go.

When you play a video on a computer, the data is transferred from one device, such as a CD, to the RAM, or memory. If there is too much data on the freeway between the CD and the RAM, the freeway gets clogged up and the video does not play smoothly. When you compress video you need to determine how much data you need to create a good image. If you have a lot of action you need a higher data rate because there are a lot of changes happening. If the video is a talking head, there are not many changes so you do not need as much data.

One setting option you can choose that will help with data rate is VBR, or variable bit rate. A **variable bit rate** means the computer looks at the video and then determines which frames of the video need more data and which frames need less. Say you cut from an interview to a motorcyclist jumping 1000 cars. The interview section does not need much data, but the motorcycle jump does. Without VBR both parts get the same amount of data. The interview has enough data to look good, but the motorcycle jump might not look very good because it does not have enough data. With VBR, the motorcycle jump gets more of the data than the interview and both sections look good.

The second setting option is the audio codec. The higher the audio data rate, the better the audio quality. But like the video, sometimes you do not really need a high audio data rate. You do not need a lot of data for a person that is just talking. However, if you have a lot of music or explosions or other things going on you want to have a higher data rate. The rule of thumb is give the audio and video just the amount of data they need to look and sound good. At some point, with both audio and video, you reach a point of diminishing returns. This means that at some point the video and audio look and sound as good as they are going to. Forcing more data per second will not make them look or sound any better and you clog up the data freeway with data that is not doing anything.

The third setting option is the frame size or resolution. If you remember, DV resolution is 720×480. You need enough data to support 345,600 pixels in each frame. If you reduce the resolution, say to 320×240, you only need enough data to support 76,800 pixels for each frame. This allows you to reduce the data rate, but still compress decent looking video. Not long ago 320×240 was about as big as you wanted to go with compressed video because larger video, like 640×480, required too much data for most computers to play the files back smoothly. Most new computers, however, can handle more data, so go ahead and experiment with the compression and make the video frame as big or small as you like.

The last setting option is the frame rate. In the past, it was a good idea to reduce the frame rate from 30 to 15 frames per second because, as with resolution, if you have half as many frames you only need half as many pixels per second of video. For example, if you have a 320×240 video at 30 fps you have to have enough data for 2,304,000 pixels each second, while at 15 frames per second you only need to support 1,152,000 pixels each second. Do not worry, the motion still looks pretty good at 15 fps as opposed to 30 fps. Today, however, you can experiment and figure out what data rate works best with what you are doing.

One question you may be asking at this point is "What about streaming?" A few years ago video streaming was the hot topic for anyone who wanted to put a video on the Internet. Everybody wanted to know who was streaming video and how they were doing it. Now, however, it is difficult to tell the difference between downloadable files and true video streaming.

With a downloadable file, a user logs on to your Web site and copies the files from your server to his or her computer's hard drive. The user then has a copy of your video. In this lesson you will create downloadable video files.

True streaming video is still a video file, but it works differently than a downloadable file. Instead of copying the file to a user's hard drive, the user simply plays the file from the server. It is like the viewer is using somebody else's hard drive. The video loads frame by frame into the RAM in the viewer's computer, and then the video is thrown away so the next frame can be loaded. The video is not saved to the user's hard drive.

> **Hot Tip**
>
> I usually do not change much in the presets. I may change the frame size or the data rate depending on what I am doing, but I do not usually alter which audio codec goes with which video codec unless I am really familiar with both codecs. Use the presets as they are and change the elements only when you really need to.

A common approach to streaming is progressive downloading. Many videos delivered over the Internet use this type of download. The idea is that the video begins copying onto the user's hard drive and starts playback when there is enough video to play the video from beginning to end without having to wait for more data to be sent. For example, a two minute video starts downloading to the viewer's computer. When one minute of the video has downloaded the computer determines that the second minute of video will download before the first minute has finished playing. At that point the video starts playing and, unless there is a network problem or some other problem, the video plays from beginning to end without the viewer noticing any stops or starts.

Note

The Internet is full of videos and video clips that average, everyday folks have uploaded. Each site, such as YouTube, or Google Video, has its own required or recommended compression formats, rates, and sizes, so it is definitely not one size fits all. Make sure you visit each site and find out what the compression requirements are to save yourself a lot of trouble.

Compression Decisions

By now you should have a basic understanding of what compression does and why you want to compress video. You also understand the elements of compression settings so you can determine how the codec compresses your video. Now, however, you need to decide how the video is seen. Will you put the video on the Internet or on a DVD? This decision helps you determine which codec to use and gives you an idea of what kind of compression is appropriate.

If you deliver your video to a DVD you don't have a choice of which codec to use because the standard for DVD delivery is MPEG-2. To create a file for delivery over the Internet, however, you first need to decide which player you require viewers to use. The two choices for compression are QuickTime Player and Windows Media Player. QuickTime Player is found for both Mac OS and Windows, but Windows Media Player is probably more common because more people use Windows machines. In the section that follows, we compress for Windows Media using Premiere Elements, and for QuickTime using Final Cut Express.

Exporting a Video as a QuickTime Movie out of FCE for Web Delivery

QuickTime is more than a simple media player. It takes more time and space than we have here to really cover how QuickTime works, so for now just remember that it is like a media player that can play VHS, Beta, Digital Beta, and DV tapes. We compress the video using H.264, which is a very common video codec that produces a high quality, but low bit rate video. The file is not very big, but the video looks nice.

STEP-BY-STEP 11.1

In this Step-by-Step you export your video as one that will be a small enough file to be burned onto a CD or delivered over the Internet. You use a preset, which means that you chose how the movie will be delivered. In this case, it will be compressed so that people with a broadband Internet connection can download and watch the movie.

1. Launch Final Cut Express and open **SBS10-8.fcp** from your data files folder. Make sure that the timeline has been selected and is the active window, otherwise you will end up compressing something other than the sequence you edited.

2. Click **File**, point to **Export**, and then click **Using QuickTime Conversion**. This opens the Save dialog box as seen in Figure 11-3.

FIGURE 11-3
Save dialog box

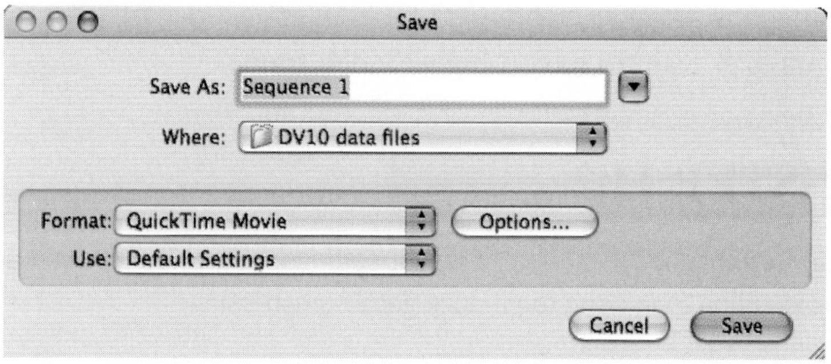

3. At the bottom of the window you see the Format drop-down box. It should read QuickTime Movie. If it does not, click the up and down arrows in the blue box and then click **QuickTime Movie**.

4. Under the Format field, you see the Use option. Click the up and down arrows in the blue box and click **Broadband-High**.

5. Click the Save as box and type **SBS11-1.mov**.

6. Choose where you want to save the movie, probably your solution folders.

STEP-BY-STEP 11.1 Continued

7. Click **Save**. A progress bar appears as the movie is compressing to the destination you chose as seen in Figure 11-4.

FIGURE 11-4
Compression progress bar

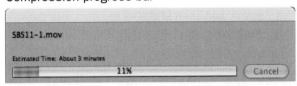

8. Leave the program open for the next Step-by-Step activity.

Sometimes, the presets do not really work for what you are trying to do with the video. For example, videos with a lot of action require a higher data rate in order for the action to be clear, or in other cases, you may want to have a bigger video frame.

S TEP-BY-STEP 11.2

This Step-by-Step activity shows you how to adjust settings, as well as give you an idea of what details you need to pay attention to in order to produce good compression.

1. With Final Cut Express open, open **SBS10-8.fcp**. Make sure that the timeline is selected and the window is active.

2. Click **File**, point to **Export**, and then click **Using QuickTime Conversion**.

3. In the Format field, make sure that QuickTime displays.

4. Click the **Options** button. This brings up the Movie Settings dialog box. On the right side of the window in the video area you see the default settings as seen in Figure 11-5.

STEP-BY-STEP 11.2 Continued

FIGURE 11-5
Movie Settings dialog box

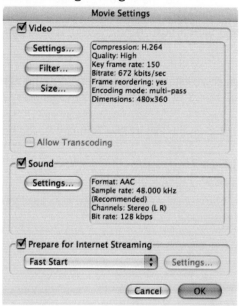

5. Click the **Settings** button to open the Standard Video Compression Settings window as seen in Figure 11-6.

FIGURE 11-6
Standard Video Compression Settings window

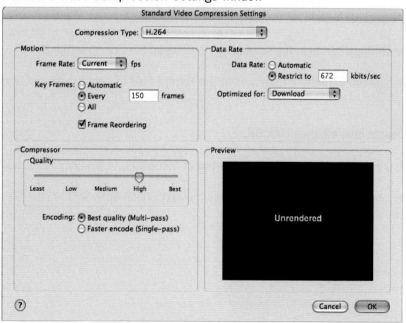

6. Click the up and down arrows next to Compression Type, then click **Sorenson Video 3**.

STEP-BY-STEP 11.2 Continued

7. Click the arrows next to the Frame Rate window and click **15**. This means that instead of the video playing back at 30 frame per second, the video plays back at 15 frames per second.

8. Click the **Automatic** radio button next to Keyframes. This means that the codec adds keyframes when needed.

9. To Adjust Compressor Quality, click the slider and drag it to the right. Stop at **High**. Click **OK** to close the window.

10. Click the **Size** button to bring up the Export Size Settings Dialog box as seen in Figure 11-7.

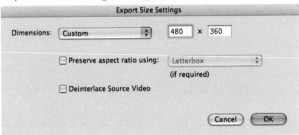

FIGURE 11-7
Export Size Settings dialog box

11. Click the blue box with the arrows next to Dimensions and then click **320X240 QVGA**.

12. Click the **Deinterlace Source Video** check box. If you remember from earlier, video images are interlaced. Computer monitors, however, are not interlaced. Removing the interlacing improves the compressed video. Click **OK**.

13. Click the **Settings** button in the sound area to bring up the Sound settings.

14. Click the blue box with the arrows next to the Format field to bring up the list of audio compression formats.

15. Click **QDesgin Music2**, then click **OK**.

16. Click the checkmark next to **Prepare for Internet Streaming** to deselect it. Click **OK**.

17. Save the project as **SBS11-2.mov** and leave the application open for the next Step-by-Step activity.

Exporting a Video as a Windows Media File out of Adobe Premiere Elements for Web Delivery

In this Step-by-Step activity, you export the video out of the timeline from Adobe Premiere and export it as a .wmv file. Remember from our earlier discussion that with a .wmv file the video plays back using Windows Media player.

STEP-BY-STEP 11.3

In this Step-by-Step, you use Premiere Elements to create a video file that plays in Windows Media Player.

1. Launch Adobe Premiere Elements and open **SBS10-14.prel**. If a dialog box appears asking for the location of the file Colonial Parade.avi, you need to browse to where the file is stored, then click **Select**.

2. Click the **Export** button above the Monitor panel as seen in Figure 11-8. A drop-down menu appears.

FIGURE 11-8
Export button and drop-down menu

3. Click **Windows Media**. This opens the Export Windows Media dialog box as seen in Figure 11-9.

FIGURE 11-9
Export Windows Media dialog box

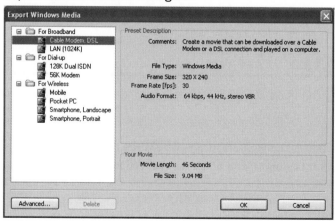

4. Preset options appear on the left side of the dialog box. Broadband and Dial-up refer to Web delivery, and Wireless settings are for cell phones and other portable devices. If the folders are not open, click the plus sign to the left of each of the folders to open them.

STEP-BY-STEP 11.3 Continued

5. Click **LAN (1024K)** under the For Broadband folder. You see the Preset Description on the right side of the window. The file type is Windows Media, the frame size is 320 × 240, the frame rate is 30 fps, and the audio will play back at 96 kbps, with 44kHz variable bit rate as seen in Figure 11-10.

FIGURE 11-10
Preset Descriptions

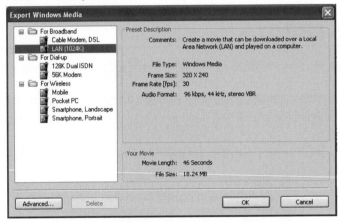

6. Click **OK**. This opens the Save File dialog box.

7. In the File name text box, type **SBS11-3**. In the Save in drop-down list, navigate to the location where you want the saved video to go.

8. Click **Save**. The Rendering dialog box appears and shows the progress of the compression. When the compression is finished, leave the project open for the next Step-by-Step activity.

Video, however, does not always fit in the presets. The delivery method (over the Internet or on a CD) plays a part in what compression is needed, as well as the amount of action, or motion in the video.

STEP-BY-STEP 11.4

In this Step-by-Step you make changes to the preset and compress the same video we compressed in the previous Step-by-Step activity.

1. With Adobe Premiere Elements open and **SBS10-14.prel** open, click the **Export** button above the Monitor panel. A drop-down menu appears. Click **Windows Media**. This opens the Export Windows Media dialog box.

2. Click **LAN (1024K)** under the For Broadband folder if necessary.

STEP-BY-STEP 11.4 Continued

3. Click the **Advanced** button at the bottom left. This opens the Export Settings dialog box so you can see and change the settings. The top window shows the current settings as seen in Figure 11-11.

FIGURE 11-11
Export Settings dialog box

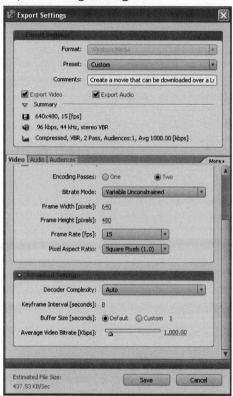

4. Click the number to the right of Frame Width[pixels] and type **640**. Click the number to right of Frame Height[pixels] and type **480**. Refer to Figure 11-11.

5. Click the arrow next to Frame Rate[fps], and then click **15**. (Refer to Figure 11-11.)

6. In the Advanced Settings section click the number next to where it says Keyframe Interval [seconds] and type **8** to change the keyframe from a keyframe every 5 seconds to a keyframe every 8 seconds.

7. Adjust the Average Video Bitrate by clicking the slider and dragging it to the right until the number to the right reads 1000. Alternately, you can click the number next to the slider and type **1000**.

STEP-BY-STEP 11.4 Continued

8. Click the Audio tab as seen in Figure 11-12.

FIGURE 11-12
Audio tab window

Audio tab window

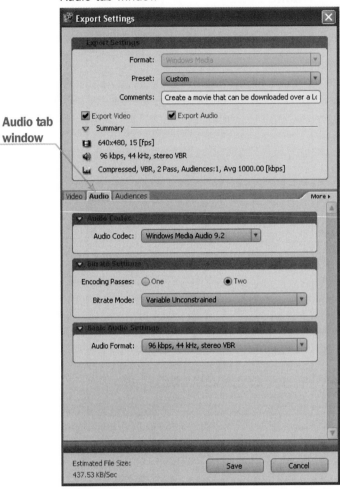

9. Click the arrow next to Audio Format, then click **128 kbps, 44kHz, stereo VBR**.

10. Click **Save**. This brings up the Choose Name dialog box. Type **My setting** and click **OK**. You do not want to replace the preset, you just want to save your own. The Export Windows Media dialog box appears and your new setting appears in a Custom folder. Click **OK**. The Save File window appears.

11. In the File name text box, type **SBS11-4** and navigate to where you want to save the file, then click **Save**.

12. After the file finishes compressing, close **SBS10-14.prel**. Do not save. Exit Adobe Premiere Elements.

Creating a DVD Using iDVD

A few years ago you needed a lot of money and specialized hardware and software to produce a DVD. Now, however, you can buy software like Adobe Premiere Elements for around a $100, or you can buy a Mac with iDVD included in the price of the Mac. With the price of a blank DVD literally just pennies, why would you do anything other than put your home movies (or the project you are creating) onto DVD?

Before we get into producing a DVD let's talk about DVDs and what they are. Some people define DVD as Digital Video Disc, while others say it means Digital Versatile Disc. A DVD is just a medium to store data. The DVDs you are most likely to purchase have a capacity of around 4.7 gigabytes, while a CD can hold up to about 800 megabytes. A single DVD, then, can hold roughly about the same amount of data as five or six CDs.

You can burn data, like word processor files, family pictures, or QuickTime and Windows Media files to a DVD and access them like you would access any data on your hard drive. If you want to make a DVD like the DVDs you watch movies on, however, you need to compress the video into MPEG-2 and burn the disc in a specific way so DVD players know how to access the data.

The nice thing about software like Premiere Elements and iDVD is that they do all of the work for you—simply add video and a little information and the software burns the disc for you.

In order to create a DVD using FCE and iDVD you need to export the video from FCE, then bring that exported video into iDVD. This is a different export than using QuickTime conversion. Exporting using QuickTime movie matches the settings you used in the edit. For example, if you have been using a DV compression in FCE, it exports using the same compression with the same data rates, etc. The video is a self-contained movie, meaning it does not have to look anywhere to find parts and pieces of itself. The video can now be taken into iDVD and made into a DVD.

> **Note**
>
> There are several different types of DVDs, such as DVD-R, DVD+R, DVD+RW, and DVD-RW. There are more as well, but you do not really need to worry about what all the initials mean right now, and explaining what the + and – means would take a while to completely explain. You need to read the owners manual or instructions for the DVD burner so you buy the right DVDs.

> **Note**
>
> Those expensive DVD authoring and compression systems I talked about earlier are still used. Many of those systems offer hardware encoding, meaning the video is captured straight to storage as MPEG-2 files as opposed to being captured into Final Cut Express and then changed from DV format into MPEG-2. The systems also offer more control over the encoding process. Anybody can learn MPEG-2 compression, but it is an art unto itself and is something that requires time, skill, and experience. People who produce the "Hollywood" DVDs are skilled professionals and take a lot of time to get everything just right.

STEP-BY-STEP 11.5

In this Step-by-Step you export the video out of FCE in a way that allows good DVD compression.

1. Launch FCE if it is not already running and open **SBS10-8.fcp**.

STEP-BY-STEP 11.5 Continued

2. If the sequence you want to put on the DVD is not already in the timeline, select it from the Browser window.

3. Click **File,** point to **Export**, and then click **QuickTime movie**.

4. In the Save dialog box, type **SBS11-5**. Make sure that **Audio and Video** appears in the **Include** field box. Make sure that **Markers** is set to **none**, and that **Make Movie Self Contained** is selected with a check in the check box.

5. In the **Where** field box, make sure that you save the file to either the Movies or iTunes folder on the MAC so iDVD can find it. Click **Save**. The progress bar appears and the file is exported to the Movies or iTunes folder. The file is done when the progress bar closes.

6. Launch iDVD. You may need to find the program icon in the dock or open the applications folder on the hard drive.

7. When the Opening screen appears, four options for projects appear. Click **Create a New Project**. This opens the Create Project dialog box.

8. In the Save As field, type **SBS11-5**. In the Where field, browse to the folder where you want to store the file and select it.

9. Make sure the Aspect ratio is set to 4:3. Click **Create**. This brings up the main creation window as seen in Figure 11-13.

FIGURE 11-13
Main Creation window

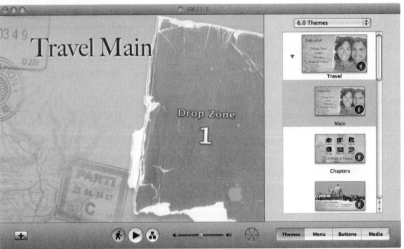

10. On the left is the Layout window where you drag and drop your video and create the menu. On the right is where you identify which themes you use, make changes to the menu, create buttons if you choose, and identify the media you add to your DVD.

STEP-BY-STEP 11.5 Continued

11. Before we move on, you want to have the computer process the video in the background while you are creating the menus. This means the original video file we exported from FCE will be copied as an MPEG-2 video file so it is compatible with the DVD specifications. Click **Advanced**, then click **Encode in Background**. If there is already a checkmark next to Encode in Background, just leave it alone.

12. Leave the file open for the next Step-by-Step activity.

Create a Simple Menu in iDVD

A few days ago I watched a movie I had on an old VHS tape. I got half way through before I fell asleep. The next day I decided to finish the movie. I tried to get to the menu, but then I realized I was watching a tape, not a DVD. I had to wait for the tape to rewind to the point where I fell asleep before I could watch the movie. Had I been watching a DVD I would have just gone to the menu, selected the scene I wanted to watch, and started watching.

A DVD menu gives you options that allow you to access specific parts of the video without having to rewind or fast forward through the stuff you do not want. The menu also can give the DVD a unified and thematic look. One of the ways you can customize the look of the menu is to use a clip of the video in a small window as a button instead of just text. Let's start off by selecting a theme. This creates the menus for the DVD and determines how the DVD looks to the viewer.

STEP-BY-STEP 11.6

In this Step-by-Step activity, you create a menu for the DVD.

1. With the file still open from Step-by-Step 11.5, click the up and down arrows in the blue box next to where it says 6.0 themes. This brings up the iDVD version theme options. In the drop-down menu, click **4.0 themes**.

STEP-BY-STEP 11.6 Continued

2. Use the scroll bar on the right and scroll down and click **Transparent Blue**. This loads the Transparent Blue theme into the Layout window as seen in Figure 11-14.

FIGURE 11-14
Transparent Blue theme

3. You are now ready to add media to the layout. Click the **Media** button in the bottom-right corner of the window. This loads the available media.

4. Click **Movies** in the upper-right corner of the screen as seen in Figure 11-15.

FIGURE 11-15
Location of the Movies folder

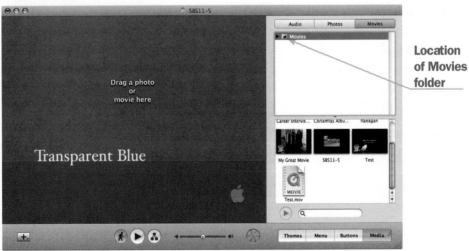

5. Click the Movies folder from the upper half of the window.

6. Click **SBS11-5** and drag it to the blue section at the bottom of the Layout window. SBS11-5 is now the video that will play.

STEP-BY-STEP 11.6 Continued

7. Click **File**, then **Save As**. Type **SBS11-6,** then click **Save** and leave iDVD open for the next Step-by-Step.

You could now burn the DVD and it would play just fine, but let's make some adjustments to it to make it better.

STEP-BY-STEP 11.7

In this Step-by-Step activity, you add chapter points to the video so viewers can skip to different parts of the video, and you also change the text in the menu button. You also adjust the motion menu video so only a few seconds of video plays instead of the entire movie.

1. With iDVD in the edit mode (not Previewmode) click **SBS11-5** in the Layout window.

2. Click **Advanced**, then click **Create Chapter Markers for Movie**. This brings up a dialog box that asks to identify how often to create chapter markers. Type **1** so you have a chapter marker every minute. Click **OK**.

3. Double-click **Transparent Blue text** in the Layout window. This highlights the text. Type **My Cool Video**.

4. Click the **Menu** button on the bottom-right side of the window as seen in Figure 11-16.

FIGURE 11-16
Menu button

5. Change the Title text font by clicking the up and down arrows next to **Hoefler Text**. This brings up a menu with the different font options. Click **Futura**.

STEP-BY-STEP 11.7 Continued

6. Change the text font size by clicking the up and down arrow next to **38.** Select **48.**

7. Click anywhere in the Layout window except in the text box. This finishes the change. If you want to go back to the original font and size, click **Reset Text** under the Font Size window.

8. Double-click **SBS11-5** in the blue area.

9. Play and click outside of the text box.

10. Click the **Media** button at the bottom-right of the window to open the media. Click **Movies** at the upper-right side and on the Movies folder underneath if you do not see your available movies.

11. Click and drag the **SBS11-5** movie from the window and drop it where it says Drag a photo or movie here. This makes the video part of the background. The video plays in a loop from beginning to end of the video.

12. Click **Menu** at the bottom right of the window.

13. At the top right side you see the Loop duration. You want the menu background to show a preview of the video, not the entire video.

14. Click the slider and slide it to the left until the little pop-up window shows 20 seconds. The loop will now play for 20 seconds.

15. Save the project as **SBS11-7** and leave iDVD open for the next Step-by-Step.

> **Note**
>
> If you use a still image instead of the video you need to place the image in the iPhoto folder before you try to access it. iDVD will not look for the image if it is not in the iPhoto folder. You access images by clicking on Photos instead of Movies at the upper-right side of the iDVD window and then dragging it to where it says Drag a photo or movie here. The background is a static image instead of the movie. If you don't want the entire movie to play in the background, you need to adjust the loop duration for the menu.

Now you are ready to preview your work and see what you have created. You do this by using the Preview window in iDVD. The Preview window allows you to test the DVD to make sure the menu takes you where you want it to take you, and make sure it plays the right video when you hit play.

STEP-BY-STEP 11.8

In this Step-by-Step activity, you preview the work you created.

1. Click the **Preview DVD playback** button at the bottom of the window (it looks like a play button). This opens the Preview window. You should see the video layout just like you created it.

2. Click **exit** on the DVD control panel as shown in Figure 11-17, or click the **x** in the upper-left side of the Preview window to close the preview.

STEP-BY-STEP 11.9 Continued

FIGURE 11-17
DVD Control panel

3. Click the **Show the DVD Map** button to the right of the Preview button. This opens the DVD map and shows the DVD navigation as seen Figure 11-18.

FIGURE 11-18
Show the DVD Map

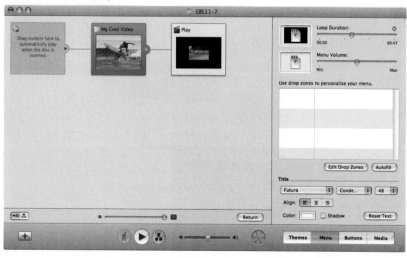

4. Click the **Show the DVD Map** button again so you can access the scenes menu. Double-click the Scene menu and repeat steps 3–7 and 10–14 from Step-by-Step 11.7 to change the menu and background.

5. Save the project as **SBS11-8** and leave iDVD open for the next Step-by-Step activity.

Burn a DVD from iDVD

Next you want to preview your DVD one more time before you burn it. If you have any issues, go back into edit mode and change what you do not like. Once you have made all the changes you want, it is time to burn the DVD.

Before you try to burn a DVD, make sure you have a DVD burner connected or installed on the computer. The computers I use have built-in DVD burners, but it is not easy burning a DVD if you do not have a DVD burner. A DVD burner writes data to a blank DVD.

STEP-BY-STEP 11.9

In this Step-by-Step activity, you burn a DVD. If you do not have a DVD burner, you cannot complete this Step-by-Step.

1. Put a blank DVD into the DVD burner. Most new Macs come with a DVD burner.

2. Click the **Burn this iDVD project to a disc** button. This is a strange looking circle below the bottom-right corner of the Layout window. The **Creating your DVD** window appears as shown in Figure 11-19. It may take a while, so be patient. A Disc Insertion dialog box appears and tells you the disc has been created. Click **Done**.

FIGURE 11-19
Creating Your DVD window

Burn this iDVD project to a disc button

Creating a DVD Using Adobe Premiere Elements

One of the advantages of using Premiere Elements is that you do not have to export the video out of the program in order to create a DVD. You simply can create your DVD in Premiere

Elements itself. You can make changes to your video if you discover problems during the DVD creation without having to switch back and forth between different software packages.

When you create a DVD in Premiere Elements you have a couple of options. The first option is to export directly to DVD without a menu. This is quick and easy and the DVD will play when you drop it into the player. You can not burn a DVD if you do not have a DVD burner. You can work through the steps, however, by saving the files to the computer without putting it on a DVD.

S TEP-BY-STEP 11.10

In this Step-by-Step activity, you create a DVD by directly exporting to a DVD without a menu. You need to have either an internal DVD burner as part of your PC or have an external DVD burner attached to your PC.

1. Launch Premiere Elements and open **SBS10-14.prel**.

2. Insert a blank DVD into the DVD burner. Again, make sure the type of DVD you are using is compatible with the burner. Skip this step if you do not have a DVD burner.

3. Click **Export**, then **To DVD**. This brings up the Burn DVD window. At the top of the menu you see the **Burn to** option with the options to burn to disc or folders as seen in Figure 11-20. If you have a DVD burner it automatically should appear in the Burner Location window. If the burner does not appear in the window, click **Rescan**, or click the arrow to the right of the burner location window and select the burner from the drop-down menu.

FIGURE 11-20
Burn DVD window

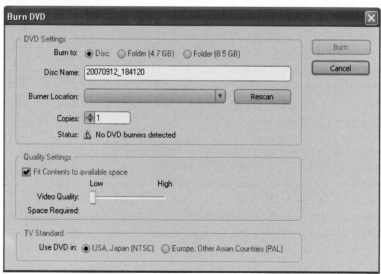

4. Click the radio button next to Disc if it is not already selected. If you do not have a DVD burner on your computer, click on Folder (4.7 GB). This will let you complete the project, but will save the files to a folder instead of burning them to a disc. If you choose to burn to a folder, you will have to identify a location on the computer to place the file.

5. Select the text in the Disc Name box and type **My cool DVD**.

STEP-BY-STEP 11.10 Continued

6. Notice the second section down, which is the Quality Settings options. The default setting is Fit Contents to available space. Below that the computer tells you how much space you need to burn the disc. You will not have any problem putting a five or ten minute piece on a DVD at the highest data rate allowed, which is 8 megabits per second.

7. Make sure that the Use DVD in option is set to USA, Japan (NTSC).

8. Click **Burn**. This brings up the Burn DVD Progress window. The window displays the current task, and the overall progress. When the DVD is finished, click **Close**. Leave the project and Adobe Premiere Elements open.

One of the beauties of DVDs is that viewers can access any section of the DVD without having to fast-forward through the entire program. In order for the viewer to do that, however, somebody has to set chapter points. Premiere Elements calls these DVD markers, and you can set them manually or have Premiere Elements set them automatically.

S TEP-BY-STEP 11.11

In this Step-by-Step you set and name DVD markers.

1. Launch Premiere Elements if it is not already running. Make sure that SBS10-14.prel is opened.

2. Move the playhead in the timeline to the point where you want your chapter point. You can move the playhead by either pressing the spacebar on the keyboard, clicking the play button in the monitor panel to play the video, or using the right and left arrow keys on the keyboard to move the playhead frame by frame.

3. Once you are at the frame where you want the first DVD marker, click the **Set DVD Marker** icon to the left of the playhead on the timeline. This brings up the DVD Marker dialog box as seen in Figure 11-21.

STEP-BY-STEP 11.11 Continued

FIGURE 11-21
DVD Marker dialog box

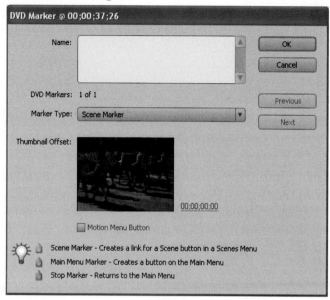

4. Click in the Name box and type **Marker 1**. You can be more descriptive if you want, but for now keep it simple.

5. Click the **Motion Menu Button** check box under the Preview window. This adds a button to the menu that will have full motion video.

6. Click **OK**.

7. Set at least one more marker by moving the playhead at least 10 seconds later in the timeline and repeating steps 4 through 6, except type **Marker 2** for the second marker.

8. Save the project as **SBS11-11.prel** and leave Adobe Premiere Elements open for the next Step-by-Step activity.

Now you are going to create the menu. Premiere Elements has a number of templates you can use, or you can import your own art work.

STEP-BY-STEP 11.12

In this Step-by-Step activity, you create a menu using Premiere Elements template.

1. With **SBS 11-11.prel** open, click the DVD icon at the upper-right corner of the window. This opens the DVD Templates dialog box as seen in Figure 11-22. This is where you create your menu and navigation layout.

FIGURE 11-22
DVD Templates dialog box

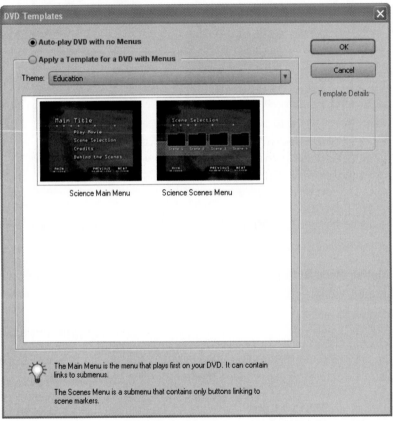

2. At the very top of the window are the options of Auto-play DVD with no Menus or Apply a Template for a DVD with Menus. Click **Apply a Template for a DVD with Menus**.

3. Click the arrow for the Theme drop-down menu and a list of types of themes appears. Click **General** from the list.

4. This brings up all of the general themes. The options show what the main menu looks like, as well as what the scenes menu looks like. Use the scroll bar to scroll down to the bottom. Click the **Wacky Main Menu** thumbnail. Click **OK** to load the wacky stuff template. You should have two lines of text, with Wacky Stuff at the top, which is the Title.

5. Double-click Wacky Stuff. This opens the Change Text dialog box. Type **My Cool Video** and click **OK**. This changes the title from Wacky Stuff to My Cool Video.

STEP-BY-STEP 11.12 Continued

6. Click the **Scenes Menu 1** at the bottom of the DVD Layout panel to load the scene selection menu. You should have a scene menu with at least two scenes, one labeled Marker 1 and another labeled Marker 2. Edit the text by double-clicking **Marker 1** to bring up the DVD marker dialog box.

7. Type **Scene 1** and click **OK**. The text is now Scene 1 instead of Marker 1.

8. Repeat steps 6 and 7 to change the text for Marker 2 to Scene 2.

9. Click the **Preview DVD** button in the bottom-right corner of the Layout panel. This shows you a preview of the DVD layout. Close the Preview window if you want to make changes to the menu, or if you are ready to burn the disc.

10. Save the project as **SBS11-12** and leave Premiere Elements open for the next Step-by-Step activity.

If you do not set DVD markers, but choose a layout from the templates, Premiere Elements reminds you that there are no markers and asks if you want the software to automatically set scene markers. You would choose yes, which brings up the Automatically Set DVD Scene Markers dialog box. This box gives you the option to set markers at each scene (Elements will decide what the scenes are), set a marker at a certain interval in minutes (such as a marker every three minutes), or simply set the number of markers. This last option even spaces the marker, so if you have an eight minute piece and tell it to create eight markers you have a marker every minute.

Burn a DVD in Adobe Premiere Elements

By now you should be ready to burn the disc. Burning a disc simply means you will write the information to a DVD and prepare it so it can be played back on a computer or set to play on a DVD player. DVDs follow a complicated standard that dictates how the data is written to the disc and then displayed during playback. Luckily, once you have gotten to this point the computer does most of the heavy lifting for you. Make sure you have a working DVD burner in your computer, or this is not going to work. If you have problems refer to the burner user's manual to make sure the burner is connected correctly.

STEP-BY-STEP 11.13

1. Launch Premiere Elements if it already is not running and make sure **SBS 11-12.prel** is open. If you did close Elements you need to load the DVD Layout Panel by clicking the DVD option in the upper-right corner of the window. If you saved your work from earlier the project is ready to burn.

2. Insert a blank DVD into the DVD burner. Again, make sure the type of DVD you are using is compatible with the burner.

3. Click the **Burn DVD** button on the lower-right corner of the DVD Layout Panel. This brings up the Burn DVD dialog box.

STEP-BY-STEP 11.13 Continued

4. Select the text in the Disc Name box and type **My Cool DVD**.

5. Click **Burn**. This brings up the Burn DVD Progress window up. The window displays the current task and the overall progress. When the DVD is finished, click **Close**.

SUMMARY

In this lesson you learned:

■ Video being delivered over the Internet or on DVD must be compressed.

■ Compression uses algorithms to save only as much data as necessary to store and display the video. These algorithms are called codecs, which is short for compression/decompression.

■ The full frames are called keyframes or I-frames. The keyframes contain all of the data necessary to recreate all video frames.

■ Delta frames are created by copying the necessary data from the previous keyframe and the other previous delta frames.

■ Video can be compressed for QuickTime and Windows Media through Final Cut Express and Adobe Premier Elements.

■ Video must be exported from Final Cut Express in order to be imported into iDVD to create a DVD. Adobe Premiere Elements allows you to create a DVD within the program itself without having to export the video.

■ Chapter points allow the viewer to navigate through the DVD at random. Menus allow you to access the chapter points and are created in both iDVD and Premiere Elements.

■ Once the DVD menu has been created and chapter points inserted, the video must be compressed to MPEG-2. MPEG-2 is the compression standard for DVDs.

VOCABULARY *Review*

Define the following terms:		
Algorithms	Delta frames	P frame
Bi-directional frame	I-frame	Spatial codec
Codec	Keyframe	Temporal codec
Compression settings	MPEG-2	Variable bit rate (VBR)

REVIEW *Questions*

TRUE/FALSE QUESTIONS

Circle T if the statement is true or F if the statement is false.

T   F   1. QuickTime is a codec.

T   F   2. Codec stands for compression/decompression.

T   F   3. Algorithm is a type of codec.

T   F   4. A keyframe is created by comparing itself to the previous frame.

T   F   5. MPEG-2 is the compression used for DVDs.

T   F   6. Keyframes are the reference point for video compression.

T   F   7. Temporal compression compares each pixel to other pixels in the same frame.

T   F   8. You can compress for both QuickTime and Windows Media Player from Premiere Elements.

MULTIPLE CHOICE

Select the best response for the following statements.

1. Which of the follow frames is created by comparing frames before and after it?
   A. keyframes
   B. bi-directional frames
   C. I-frames
   D. delta frames

2. Which file extension identifies QuickTime files?
   A. .wmv
   B. .qtf
   C. .pdf
   D. .mov

3. What does a chapter marker do?
   A. Identify the beginning and end of the video
   B. Used to start and stop the video
   C. Used in QuickTime movies to identify keyframes
   D. Used to navigate through a DVD

4. Which compression setting changes the amount of information that is used to display the video?
   A. frame rate
   B. frame size
   C. data rate
   D. format

5. Which export choice in Final Cut Express do you choose to compress a QuickTime movie to play on the Internet?
   A. QuickTime Conversion
   B. QuickTime movie
   C. Export QuickTime
   D. Export .mov

6. How do you add chapter points to a video in Premiere Element?
   A. Select Timeline, then Add chapter points from the menu bar
   B. Click the Marker button in the monitor panel
   C. Select Edit, then Markers, then Add chapter points from the menu bar
   D. Click the DVD Marker icon to the left of the timeline

7. Which export choice in Final Cut Express do you choose to compress a QuickTime movie in order to take it into iDVD?
   A. QuickTime Conversion
   B. QuickTime movie
   C. Export QuickTime
   D. Export .mov

8. Bi-direction frames reference
   A. other bi-directional frames only
   B. the previous keyframe and the following B frame
   C. the previous I-frame only or the following P frame only
   D. the previous I-frame and the following P frame

WRITTEN QUESTIONS

Write a brief answer to the following questions.

1. Explain the basic idea of compression and how it is done.

2. Explain variable bit rate and why you would use it.

3. Explain how video streaming works.

PROJECTS

PROJECT 11-1

Create a DVD using either iDVD or Premiere Elements. Include at least three chapter points, and a menu of your choosing. The video you use should be the video you shot and edited. If you do not have your own video, create a DVD using the video provided with the lesson, but create at least three chapter markers and use a different menu.

PROJECT 11-2

Visit YouTube or Google Video on the Web. Find out what their required or recommended compression formats, rates, and sizes are and compress a video to deliver through these services.

 TEAMWORK ACTIVITY

Put together a group of at least four different classmates. Have each classmate compress the same video clip on his or her own computer, then compare the results. Everybody can use a different compression, different data rate, and different file format (QuickTime or Windows Media). Compare the results. Which ones look better? Why do they look better or worse? Compare notes and recompress the videos, changing the setting for each group member in an attempt to improve the result. Did the improvements help?

 WEB PROJECT

Research video compression on the Web using search terms such as video codecs, QuickTime, and Windows Media. Research QuickTime and Windows Media wrappers as well and find out as much as you can about how they work. Write at least two pages explaining the difference between wrappers and codecs and how they affect video compression.

CRITICAL*Thinking*

ACTIVITY 11-1

Compress a single video using the same codec at least four different ways. Use different frame sizes, different frame rates, different data rates, etc. Compare the videos and then figure out how they look different, and why they look different.

GLOSSARY

A

Active sentence A sentence that follows the Subject-Verb-Object formula. Example: The boy kicked the ball.

Algorithms Instructions for compressing and decompressing video.

Amplitude The power of the wave.

Analog Replication of sound waves or light waves that you hear and see.

Antagonist The character that opposes the protagonist.

Arc light Specialized lights that require a generator or a large power source. They are typically used in big Hollywood shoots.

Audio levels The volume of the clip.

Audio operator The crew member who sets up the audio connections and makes sure the audio works.

Audio sweetening The process of adding music and sound effects to the video as well as cleaning up and improving the audio.

Audio synchronization The process of matching the audio to the action backlight.

B

Balanced audio A method for audio connection and transfer that uses a third wire to eliminate audio noise, hiss, and hum.

Bandwidth Transmission capacity of data.

Barn doors Hinged shutters that can be moved into place over the light.

Bi-directional frame Video frames that track changes. They are created by comparing pixels from the previous key frame and to the delta frames following it.

Bi-directional pattern microphone A microphone capable of a pick-up pattern where it is sensitive to sounds coming from the front and behind but not from the sides.

Bit depth The number of levels (how high and how low) a sample can recreate.

Boom microphone A microphone attached to a boom pole (or fish pole) that allows the audio operator to place the microphone close to the subject or subjects without using a stand or having somebody hold a mic.

Butterfly A large frame on which you can attach silks or scrims, allowing the user to control light coming down onto the subject.

C

C47 Wooden clothespin.

Camera A device that captures light information and translates it into a video image.

Camera operator The crew members who run the camera by moving and adjusting the shot according to the directions of the director of photography.

Candelas A measurement of light intensity.

Capture card Hardware that converts the video into digital information the computer can work with.

Capture Recording footage to a hard drive.

Cardioid pattern microphone A microphone capable of a pick-up pattern where it is sensitive to sounds from in front of the microphone but not from behind.

Cast Characters with speaking parts.

Casting The process of identifying actors that best fit the role through auditions.

Casting call Tryouts for an acting job.

Cathode ray tube (CRT) A vaccum tube that produces cathode rays from electron guns. These guns light up the red, green, and blue pixels of an image on a screen. Televisions and computer screens use this technology.

Charge-coupled device (CCD) The imagaing device in a video camera.

Chrominance The amount of saturation.

Codec Short for compression/decompression. Algorithms, or instructions, for digital media compression.

Color sampling A way of compressing video by referring to the amount of information used to describe the color in a video image.

Color temperature A characteristic of visible light measured in kelvins.

Compressed To be made smaller.

Compression settings Define the codec to be used for audio and video compression. Include the data rate for audio and video, frame size, and other compression elements.

Condenser microphone A microphone that is more sensitive than dynamic microphones but is not as rugged and requires a power source. Usually found in studio settings.

Conflict What goes wrong in a story.

Content What is written in a story.

Continuity When each shot from every setup matches in action and look.

Convex Refers to a lens type that has thin edges and a wide center. The lens focuses all of the light that hits its surface and focuses it onto the imaging device.

C-stand A stand that holds a light control in position.

Cucaloris or cookie A piece of opaque material with shapes or patterns cut out.

Cutaway Shot of something extra in the scene, such as a close-up of a person stirring a drink.

D

Delta frame The non-key frame compression frames that create themselves based on the data from the previous frame.

Depth of field The area of the image that is in clear focus.

Device control Enables the computer software to control the playback, rewind, fast-forward, and other functions of a playback deck or video camera.

Diffusion A softening of the light and the shadows the light creates.

Digital zoom Electronic zoom technique in which the pixels in center of the image are enlarged.

Dimmer Reduces the flow of electricity to a bulb.

Direction Determines the position of a shadow depending on the light source.

Director The crew member responsible for the artistic side of the production.

Director of photography The crew member responsible for making the scene look and feel the way the director wants through composition and lighting.

Documentary A nonfiction film or video.

Dynamic microphone Rugged microphones that are good for location sound. They do not require a power source (good), but they are not as sensitive as condenser or ribbon microphones (bad).

E

Editing The process of combining footage that tells your story, and eliminating the footage you do not want.

Establishing shot Shot that establishes where the character is and what is going on around him or her.

Exposure The amount of light that reaches the imaging device.

External conflict A conflict that occurs outside of the character, such as the antagonist trying to prevent the protagonist from achieving his or her goal.

Extras Actors who do not portray main characters and usually do not speak lines.

F

Field In scanning, half of a frame.

Field mixer A portable audio mixer that runs on batteries so you can adjust the microphone audio levels.

Field of view How much of the subject that is seen in the image.

Fill light One of the lights in three-point lightning that fills in the shadows created by the key light while not creating any of its own shadows.

Firewire Enterface standard for connecting devices, such as cameras and hard drives, to a computer. Also known as IEEE 1394, DV, and Sony iLink.

Focal length The distance between the lens and the imaging device when the subject is in focus.

Focal point The point at which the light passing through a convex lens comes together.

Foot-candle A measurement of light intensity.

Frame A single complete picture.

Frame-accurate editing Editing that allows the editor to identify the specific video frames that will be included, or not included, in the final edit.

Framing A composition technique in which a frame is created inside a larger video frame.

Frequency The number of times a wavelength repeats in one second.

Fresnel An enclosed spotlight.

G

Gaffer The electrician.

Gel A film that can be placed over a light source to reduce the intensity or change the color of the light. It works with the heat produced from the lights without burning or melting.

Gobo Similar to a cookie in that it shapes light.

Graphics operator A crew member responsible for creating graphics, running the graphics computer, and making sure the correct graphic is ready and is being shown.

Grip The crew member who moves things around and puts them where they need to be.

H

Halogen A substance inside of a Tungsten filament that shows the destruction that happens because of heat.

Hard light Light that produces hard shadows.

HD Stands for high definition. This uses 720 progressively scanned lines that add detail.

HDV An MPEG-2–compressed high-definition video format.

Head room The amount of space in the frame above the subject.

Hue Color information that allows a color to be classified as red, blue, yellow, orange, and so forth.

Hypercardioid A pick-up pattern similar to the supercardioid, but with a narrower pick-up pattern in the front and a slightly larger pick-up pattern in the rear. This means it will eliminate even more sounds coming from the sides of the subject.

I

I-frame A key frame in MPEG-2 compression.

Imaging device The part of the camera that receives the light from the lens and creates the video image.

In point The first frame of a trimmed clip.

Intensity The strength of light.

Interlaced video Video that uses two fields to create a complete image.

Internal conflict A conflict that occurs inside the character, such as a police officer who cannot make an arrest because of a mistake that he or she made in the past.

K

Kelvins A measurement of heat.

Key frame A complete frame. Delta and bi-directional frames are created based on the data in key frames.

Key light The main source of light. It creates the main shadows on the subject.

L

Lavalier microphone (lav) A small microphone that can be attached to the subject's clothes (or to the subject).

Lens The part of a camera that gathers and focuses light.

"The line" An imaginary line that determines where cameras should be placed to keep objects and characters in the same relative position from shot to shot in a sequence.

Linear video Video that does not allow random access to any time point or frame.

Luminance The brightness or amount of light in an image.

Lux A measurement of light intensity.

M

Master shot *See* Establishing shot.

Matching time code When the time code from each camera is the same at the same point of the program.

Mini connector Most common audio output on consumer or prosumer DV cameras.

MPEG-2 The codec used for DVDs.

N

Narrative A fictional story.

National Television System Committee (NTSC) The standard for American video. It is composed of 525 scan lines.

Neutral density (or ND) gel A gel made specifically to work with the heat produced from lights without burning or melting.

Noise Unwanted glitches in an analog audio and video signals. It shows up as "snow" or static in a video image.

Nonlinear Does not follow a straight line or a sequential order.

Nose room The amount of space in the frame in front of the subject.

O

Omnidirectional microphone A microphone that is pattern-sensitive to sounds that come from all around it.

Optical zoom Zooming in on the subject using the lens elements.

Out point The final frame of a trimmed clip.

Overexposure When more light than is needed reaches the imaging device.

Overhead *See* Butterflies.

P

P-frame Short for predictor frame. Delta frames in MPEG-2 compression.

Pan A camera movement in which the camera rotates from side to side.

Passive sentence A sentence that does not follow the Subject-Verb-Object structure. Usually follows an Object-Verb structure. Example: The ball was kicked by the boy.

Persistence of vision A theory that states that the human eye holds each still image for a fraction of a second with the image remaining on the retina long enough to blend the image with the next image.

Pick-up pattern The description of the area around a microphone where it is sensitive to sound.

Pixel (From **picture element**.) The smallest sample of a graphic image or a single point in an image.

Program A completed live shoot—what the audience sees.

Progressive scanning Creates a frame by scanning the scan lines in order 1, 2, 3, 4, and so on instead of displaying a frame in two fields.

Props Any tangible item that is handled by a character.

Protagonist The main character in a story. Also known as the hero of the story.

R

RCA connector A common, unbalanced audio connector.

Reaction shot A shot that shows the listener's reaction to what the speaker is saying.

Real-time transition A transition that does not require rendering.

Reflector Used to reflect light back onto a subject.

Rendering Processing video so that it can play correctly and smoothly.

Resolution Number of pixels in an image that determines the amount of detail in the image.

RGB Stands for Red, Green, and Blue.

Ribbon microphone A high-quality, but not rugged, microphone. Usually found in studio settings.

S

Sample rate The number of times in a second that a sample is chosen from the analog signal.

Sampling A digital method of obtaining an analog signal by picking specific points along a wave signal.

Scene breakdown sheet A written document that identifies the individual elements needed to shoot each scene.

Scene numbers Numbers that identifiy specific scenes.

Scrim A metal screen placed in front of the light to reduce the intensity without changing the hardness of the light.

Script A blueprint for a film or video.

Script breakdown The guide for production.

Sequence A series of individual shots that when edited together create a scene.

Shooting script A script that includes scene numbers and revisions.

Shot The time between when recording starts and stops. It is the basic building block of video.

Silk A piece of artificial silk that allows some, but not all, of the light onto an object.

Soft light A light that creates soft, less-defined shadows.

Sound effects/music Specific music playing in the scene, not the film score, or any extra sounds, like a door slamming.

Spatial compression Video compression created by comparing pixels to other pixels in the frame.

Special effects Graphic effects usually created by computer-aided design or physical modeling, like a spaceship moving through a galaxy.

Special equipment Specialized machines or equipment needed to physically shoot a scene, like a dolly or crane.

Story Retelling of events.

Storyboard Visual guide for the shoot. It resembles a comic book.

Strip board Consists of two parts: the strip board that identifies the information and holds the strips and the strips themselves. The strip board has the elements needed to shoot each scene.

Stunts Any kind of dangerous physical action that requires special safety precautions.

Style How a story is written.

Supercardioid A pick-up pattern in the front that is narrower than a cardioid and that picks up some sound from behind the microphone.

Switcher Equipment that accepts the signal from any number of sources and determines which shot is edited into the show and when.

T

Tape operator A crew member responsible for starting and stopping VTRs and tape playback.

Technical director A crew member responsible for running the switcher.

Temporal compression Compression based on comparing pixels between frames.

The rule of thirds A guide for image composition in which the frame is divided by an imaginary tic-tac-toe board. The subject should be placed in the area where the horizontal lines intersect the vertical lines.

Tilt Camera movement in which the camera moves up and down.

Time code A unique number based on hours, minutes, seconds, and frames that identifies each frame of video.

Trim or trimming Getting rid of the parts of a clip that are not needed in the final edit.

Tungsten The most commonly used light. It uses a metallic element to create filaments that generate light through the production of heat.

U

Underexposure When not enough light reaches the imaging device.

USB (Universal Serial Bus) Interface standard for connecting devices, like cameras and hard drives, to a computer.

V

VBR (variable bit rate) Means that the computer looks at the video and determines which parts of the video need more data and which parts need less.

Vehicles/animals Vehicles and animals that are in a scene, like a car or a dog.

VTR Video tape recorder.

W

Wattage or watt Measures the number of electrons that are moving down a wire to a light fixture or bulb.

Wavelength The distance between equivalent points on consecutive phases of a wave pattern; the length of the wave.

X

XLR A common professional-balanced audio connector.

INDEX